3-D Geometric Forms For Art

50+ Polyhedric Models & Lessons to Copy

First Edition, Soft Cover

Compiled & written by Eric Gibbons

ISBN 13: 978-1-940290-78-2

AF808909

Printer: Amazon
Publisher: Firehouse Publications
Credit: Special thank you to Polyhedra.net where many models for this book came from and more can be downloaded for free.

Limited copy rights assigned to
Printed Name of purchaser:

____

Signature of purchaser:

____

Date of Purchase: ___/___/_______

Venue of purchase:

____

Table Of Contents:

Introduction: I began to learn Origami at the age of five, quickly encountering shapes and forms. Because I had been introduced to these concepts early on, I was able to quickly grasp concepts of Geometry and Trigonometry before my peers. Research shows that students who take art through high school out perform their non-art-involved peers by over 100 points on their S.A.T exams. There is a whole genre or Origami called Units and Modular Origami that tie the ancient Japanese artform with geometry. Books on that would make wonderful companions to this resource. I have compiled the models here to be easily photocopied and used by students, artists, and those interested in the topic. Additional forms can be found for free at www.polyhedria.net. Another valuable resource can be found at polyhedra.tessera.li. There you can explore forms and manipulate them virtually.

The first few pages of this book will cover some basic lessons I have used in my classroom that tie into polyhedric forms. I have tried to organize the book with simple models in the beginning and more challenging ones in the back. When I make copies, I find it helpful to cut the binding of this book to make photocopying easier. Many machines will allow you to make enlargements, so you can scale up images to make larger forms. This book is formatted to be 8.5 x 11 inches, standard for photo copiers.

Bring geometry and art together with these fun explorations for middle and high school students. With over 50 models you can photocopy or scan, this STEM-based resource can be used for lesson planning or sub plans to keep students engaged and deepen their understanding of math, geometry, measuring, proportion, tool use, and more!

Got Art?!
Grades 4 to 12

This lesson is based on the "Got Milk" advertising campaign.

A great way for students to experience art history is to create a work of art inspired by a particular artist. My "If Picasso" book series found at www.firehousepublications.com does this with animals, landmarks, and Christmas Trees in the style of artists throughout history. They can be a fun resource for this project to help students understand the idea of emulation.

I was inspired for this project by the work of my independent student who started drawing and painting on her used lunch trays. I thought her work was cool, and the students seemed to really appreciate her work. I paired her concept with my annual art history unit for this fun mash-up. The reaction of both students and staff over our display was enthusiastic.

Our first step was to choose artists with unique visual styles. I posted a list with some artists to choose from, but students were able to select artists not on the list if their work was in a museum somewhere. Some students chose artists tied to their own cultural background. They know this will positively impact their grade when the rubric calls for making personal connections to our work. At the end of the assignment, students researched three interesting facts about their artist to display with the finished tray.

I allowed students to interpret the assignment broadly. They could use the tray "as if" if was the canvas, or they could choose to make some items 3D. We had acrylic paste available for textures, thick water color paper, and even plaster mâché if it was needed. While most worked in acrylics, our Basquiat tray was spray painted and drawn on. Some students emulated the work of the artist, others created a kind of parody.

I found a template on the internet for our milk cartons and printed it onto index paper. (Above) While some students washed out an actual carton from the lunchroom, most created on the back of the template which was cut, folded, and glued together. A few students used paper cups instead and one used a juice can for her James Rizzi-style tray. She did her meal as food from India to make a personal connection between her own culture and the artist to exceed expectations!

For students who finished early, they were asked to do extra milk cartons based on artists not yet selected. They found this to be fun and kept them actively engaged in the assignment while others continued on their trays. We had enough extra milk containers that we made a mini "Moo-Seum" display for our school atrium that was a real hit!

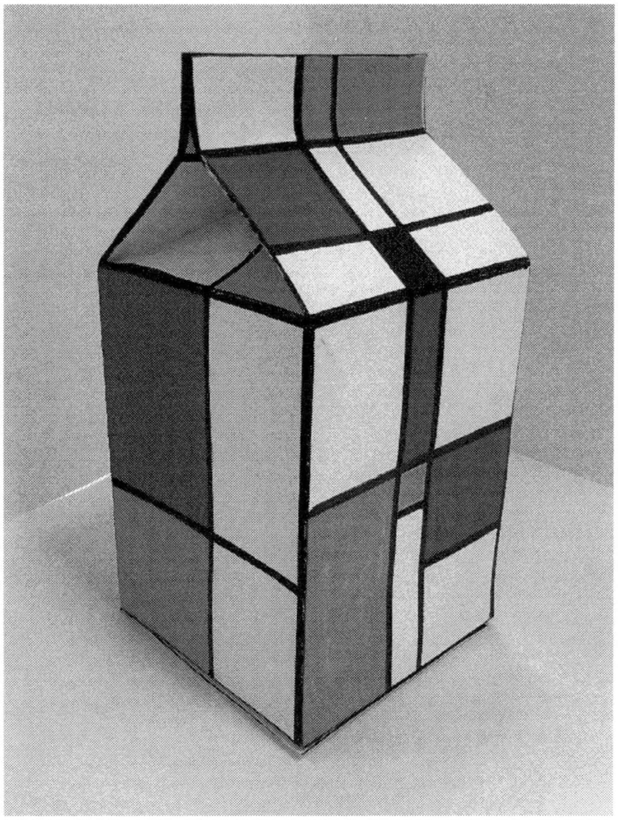

Mondrian Style Milk Carton

GOT ART?

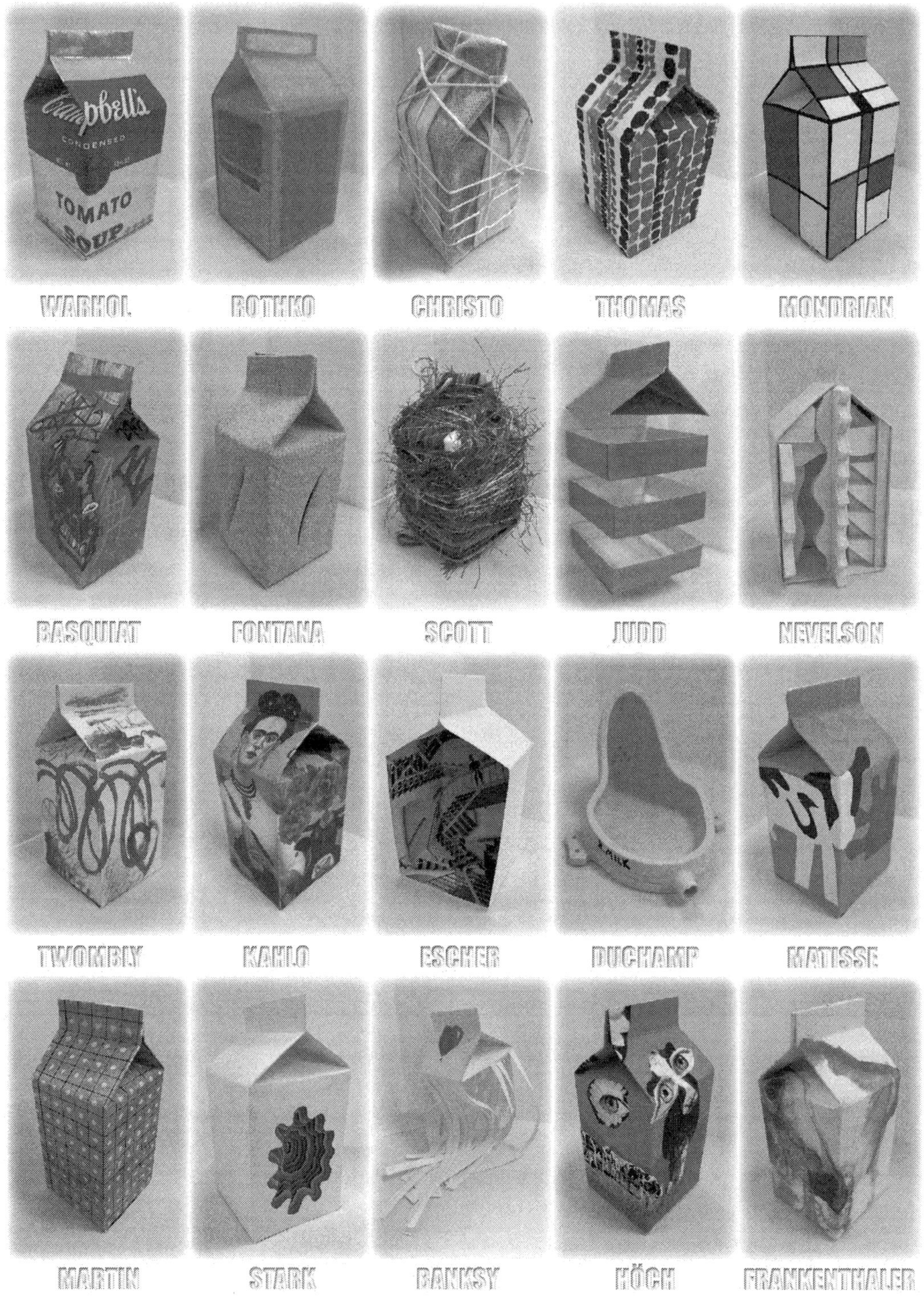

THESE ARE EMULATIONS & PARODIES; NOT THE ACTUAL WORKS OF THE NAMED ARTISTS.

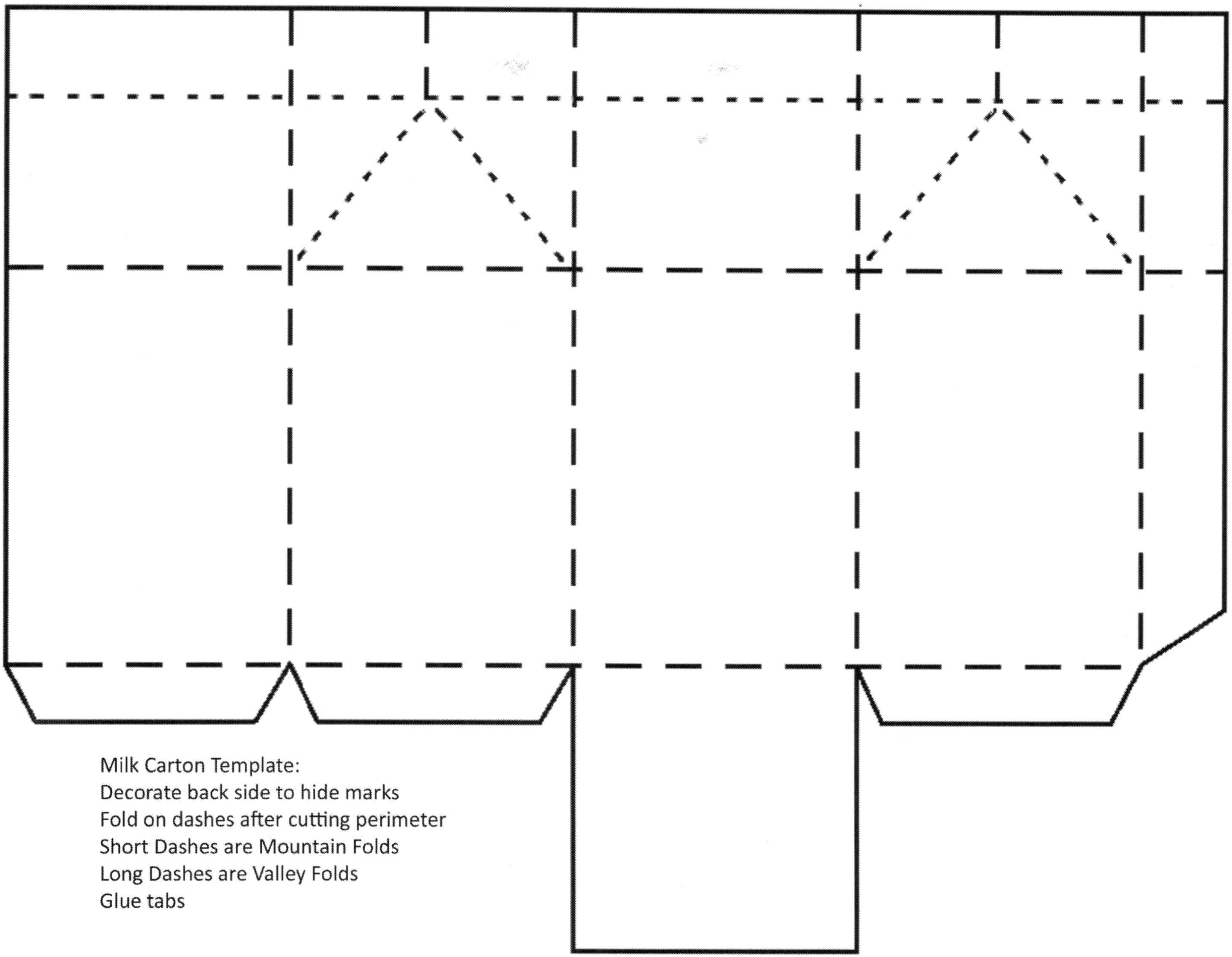

Milk Carton Template:
Decorate back side to hide marks
Fold on dashes after cutting perimeter
Short Dashes are Mountain Folds
Long Dashes are Valley Folds
Glue tabs

Foil Forms
Middle and High School

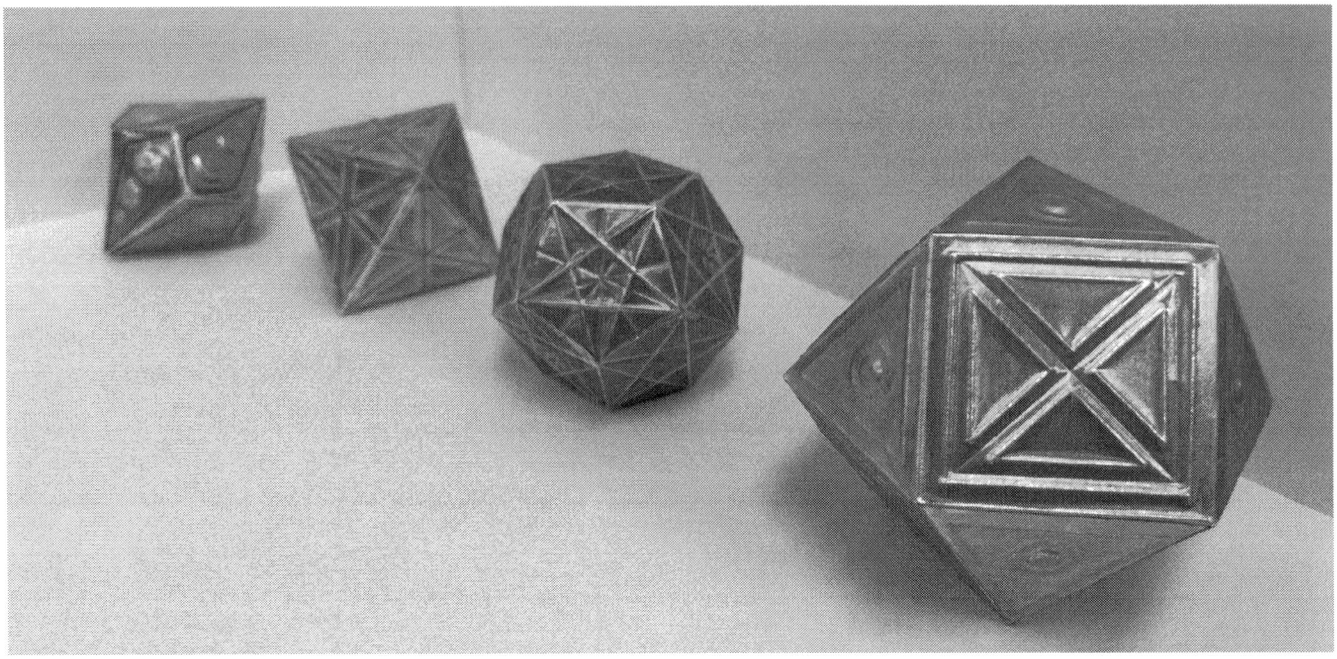

Using a thick tooling foil found at most art suppliers, we created these decorative repousse forms. We created these during the holidays, so students who wanted to use them as decorations were able to add a hook to them so it could be hung in a window or on a tree.

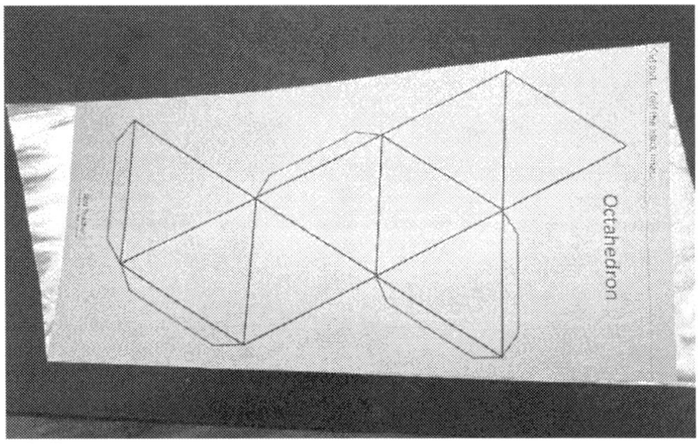

We photocopied form templates and spray glued them onto the back of our foil. They were cut out with razor blades, but scissors would have worked too. If tooling foil is unavailable, regular foil will work but photocopy onto thicker index or card stock for durability.

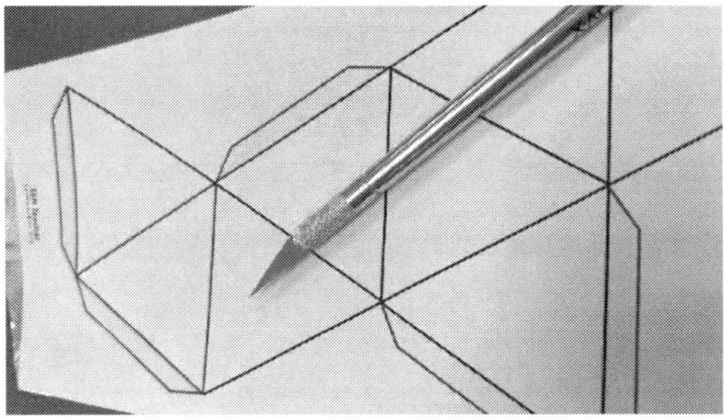

Once the forms and tabs are cut out, we traced the shapes with a ball-point pen and straight-edge on the paper side. It is helpful to work on a surface with a little "give," like a pile of paper, book cover, or linoleum block. This way the lines are more easily transferred to the foil. If the surface is too hard, like a solid counter top, lines may not transfer. The softer the surface, the more pronounced the lines will be.

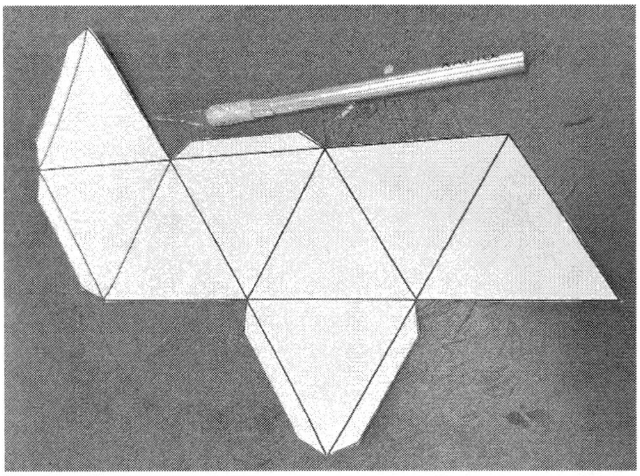

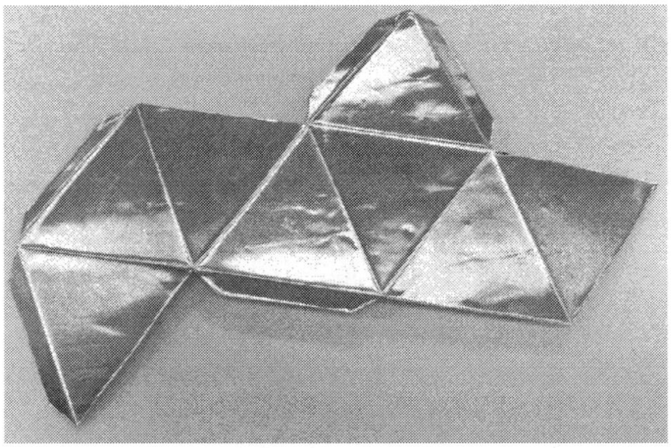

After shapes are traced, we colored in shapes with permanent markers. Sharpie brand worked well.

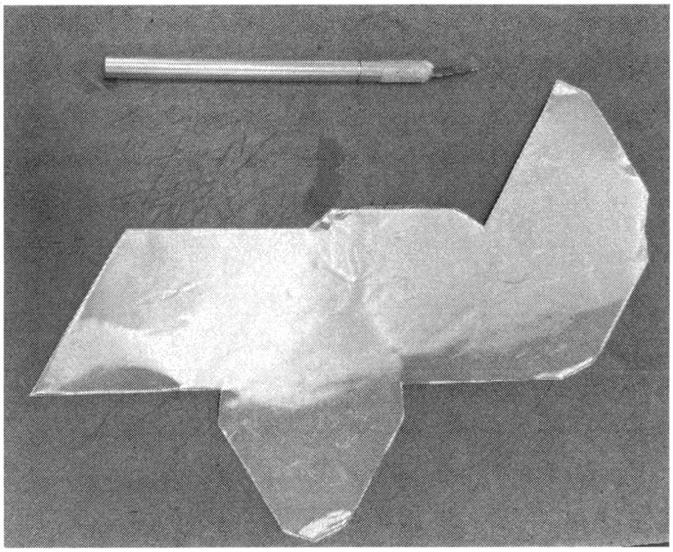

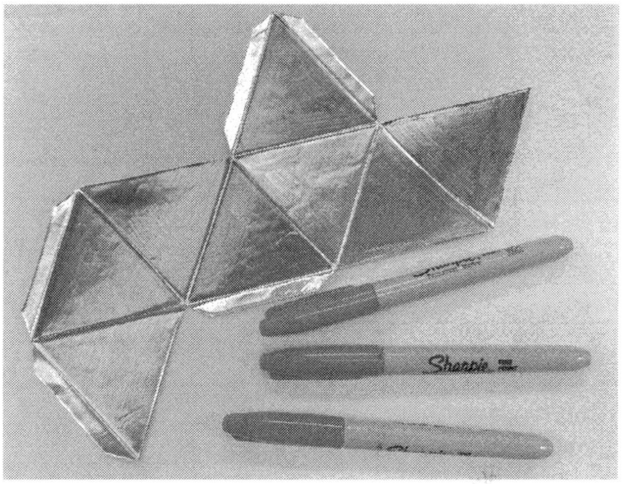

When the color was dry, we embossed the foil from both the back and front with ball-point pens. Even empty pens work well. It is important to use a straight edge for precision. You can see in the image I worked on a soft linoleum block so that the surface had enough 'give" for the lines to transfer through.

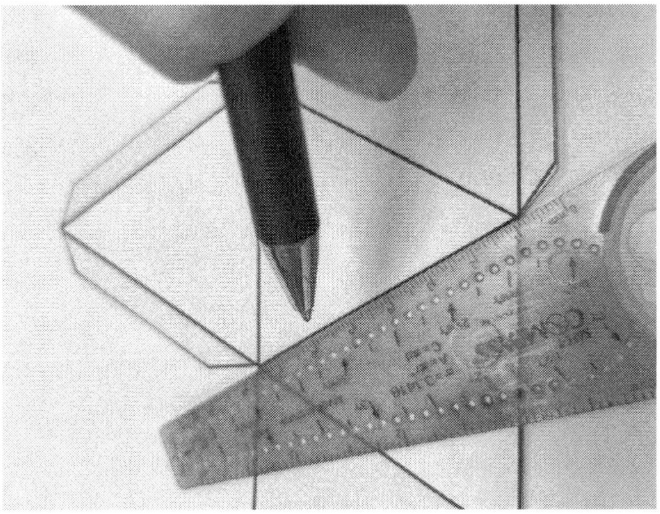

Contact cement was used to glue the forms together, though double-sided tape could be used as well. Follow directions on contact cement for good adhesion. The Elmer's Brand glue did not work well, but "Master Cemento" did. I found that on Amazon. It has strong fumes so ventilation is very important. We poured it into Elmer's bottles because they had a good applicator brush.

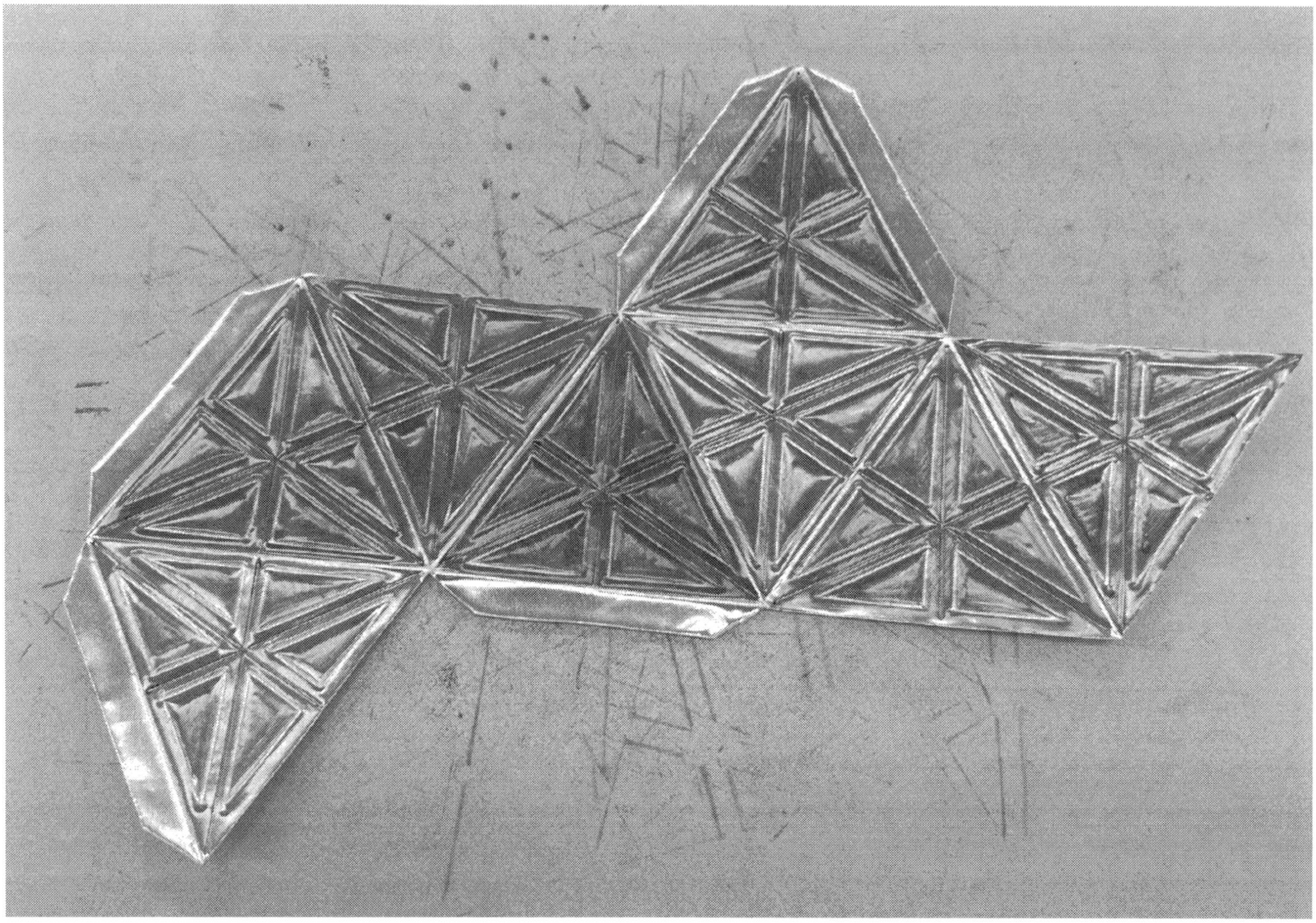

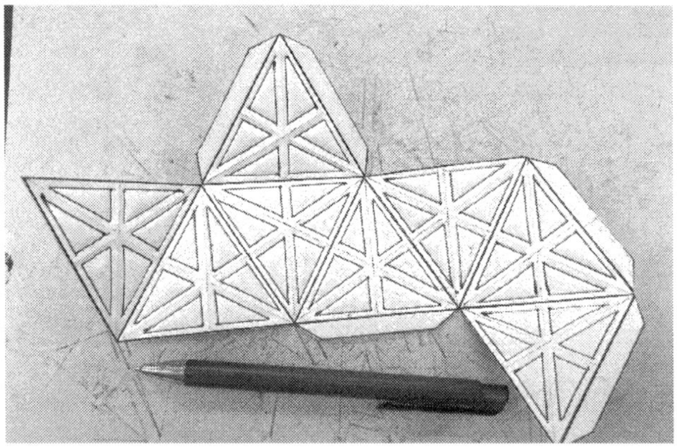

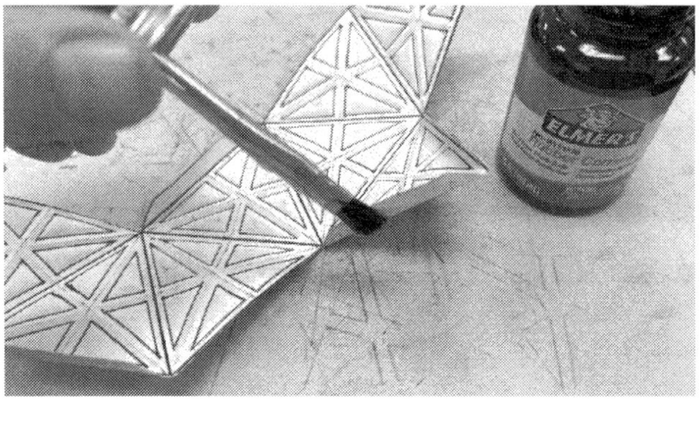

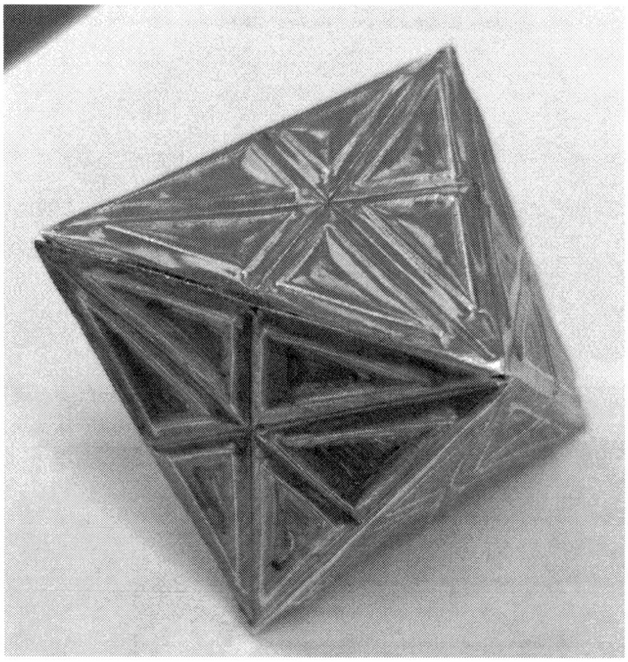

Video Tutorial: www.youtu.be/ZNDc6eRSl8k
(Lower case "L" before the 8)

Contorted Cartoons
Grades 4 and up.

Online Resources:
www.artedguru.com/home/contorted-cartoons

"Let's shade some forms!" (YaWn!)

We've seen it before, the age old exercise of shading cones, cylinders, spheres, and cubes, and boring the artistic spirit out of our students. My twist is to have students chat and write about their favorite cartoons they lived as younger kids. We laughed at how creepy Teletubbies were, or how funny SpongeBob and Patrick were, or the game characters they loved. Each student printed out an image of their favorite about the size of their open hand.

These were glued to the back of form templates I found on Google Images and had printed onto card-stock paper. We used markers and colored pencils to give them basic colors before cutting out the forms and gluing them. It was very helpful to have students pre-fold their templates with the edge of a ruler. We spoke about how folding a 2D image onto a 3D surface fractured the images in a way that Cubism also did in art history, but also how the Pop Artists of the 1960s would use cartoon images. This meant our project was a kind of modern mash-up.

Each student was to make one form, and to work in groups of 3 or 4 to draw a collection of forms as a kind of still-life drawing. (Those who finished quickly were told to sketch their individual form as practice in the back of their workbooks, like above.) The forms were placed on a white sheet of paper so that their shadows would be more evident. We started with the light on, then when we got to shadows, I turned off the lights and used the windows as out light source so that shadows would be more pronounced.

I created a video to help students through the process.

- Draw the forms and the paper they were sitting on
- Draw the cartoon so it looked like it was on a 3D surface
- Add shadows to the forms and then the cast shadows.

The cartoons helped personalize their work. By working with friends, they learned something about each other. They saw that forms change as the surface bends or turns. They learned how to shade fairly well with a project that was not too overwhelming. And they learned their teacher is pretty old and liked cartoons they never heard of.

The whole exploration took about 5 class days from printing images to final shading. We worked on 12 x 18 inch paper, but any size will do. We used cartoons, but any image on the surface would be fine, maybe even patterned paper. I just found that cartoons were a great motivator, made the lesson a bit more fun, and resulted in some fun conversations. I always try to connect student work back to their interests.

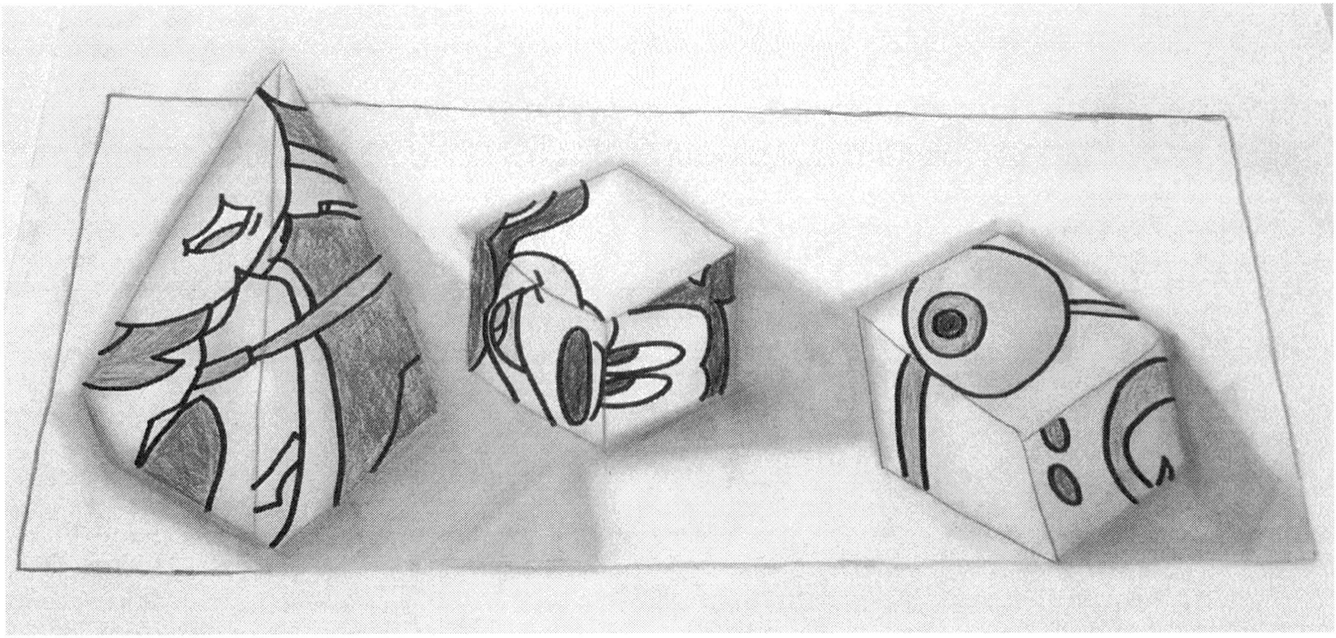

Student Sample

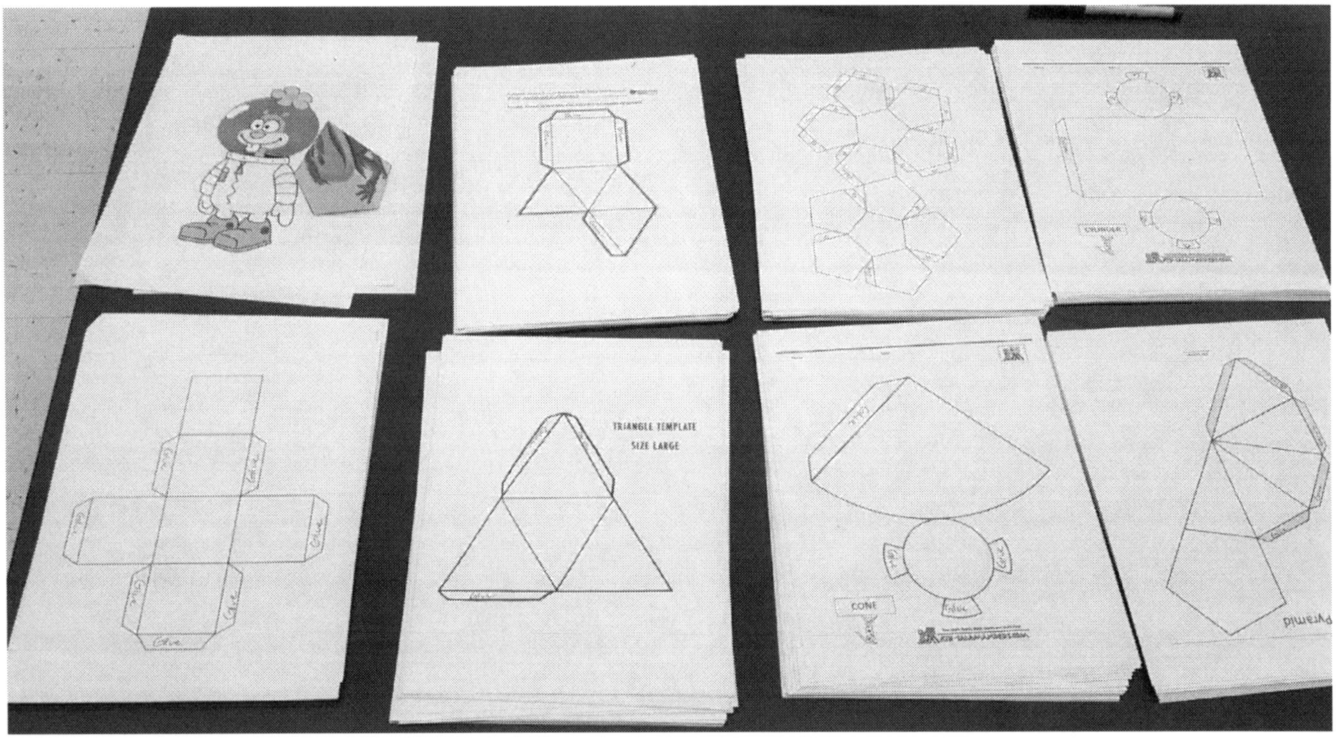

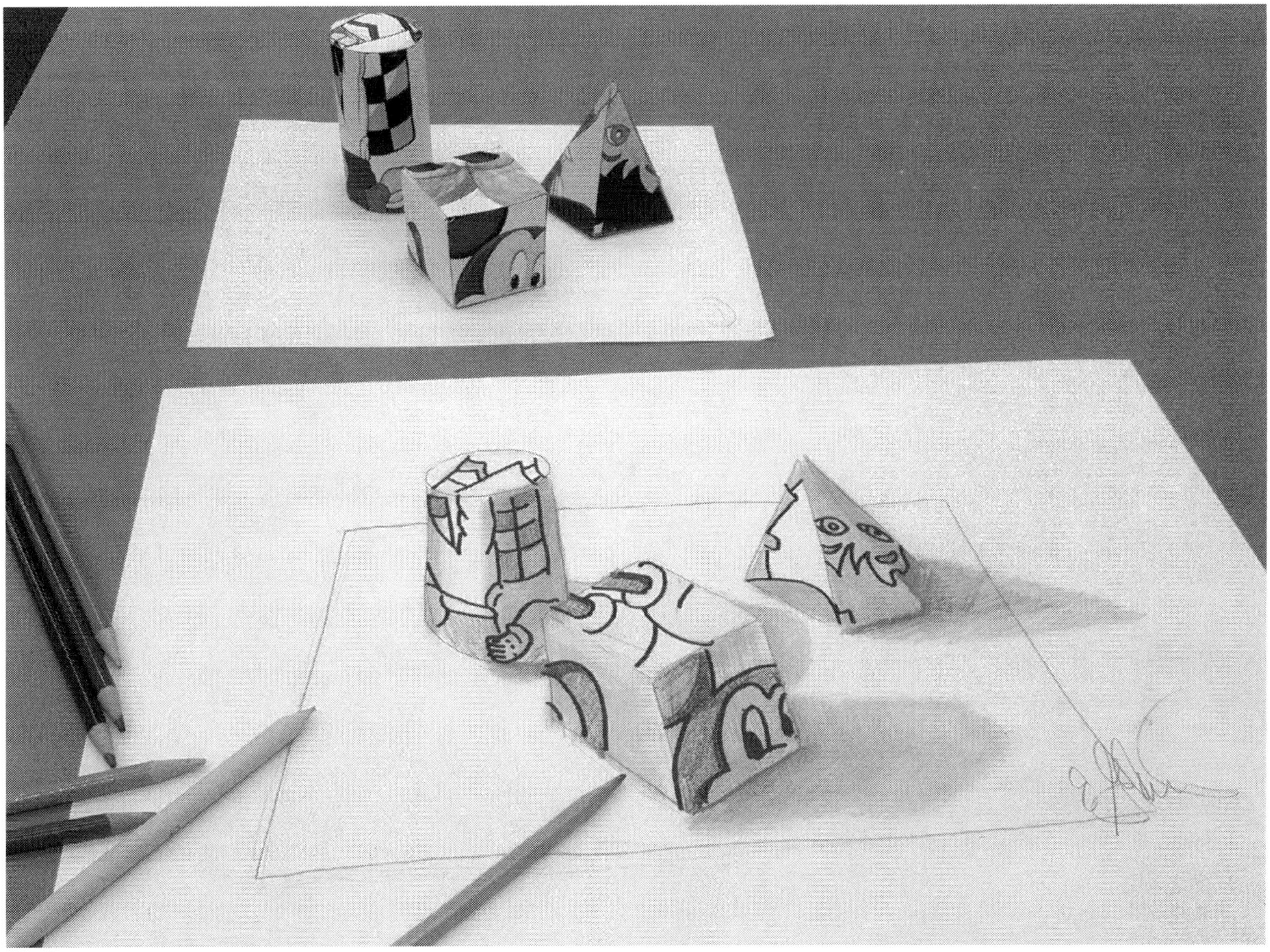

Teacher Sample

Ceramic Form Containers:

Using plaster or clay students could create containers to hide their personal treasures or even design piggy banks. The form could be expressive of the treasure or thing they would like to save up for so there is a personal connection.

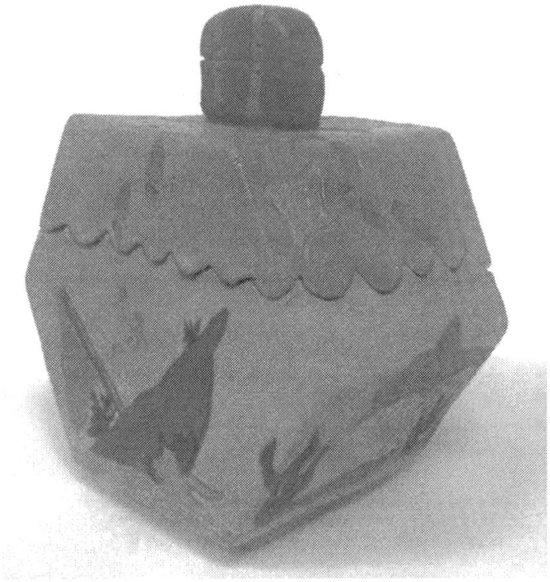

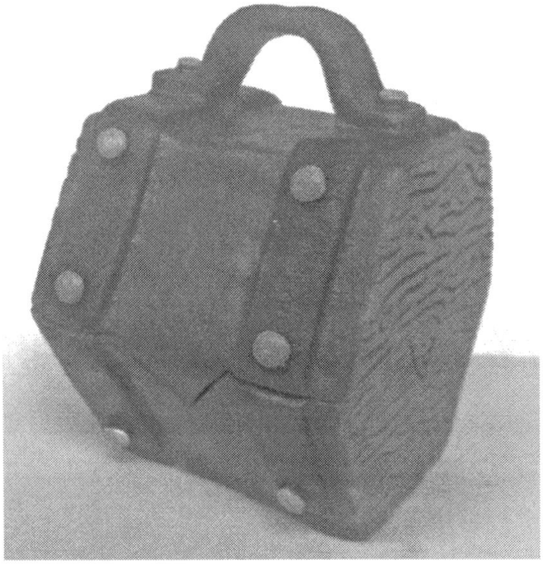

Credit Artwork by students of DeAnn Hanisch, Art teacher at Omaha Burke High School

Potion or Perfume bottle

Small ceramic forms could be turned into perfume bottles or "magic" potion containers. The form or decoration can allude to the function of the potion or the scent of the perfume. This is also an opportunity to explore package design and illustration if they make labels for their containers.

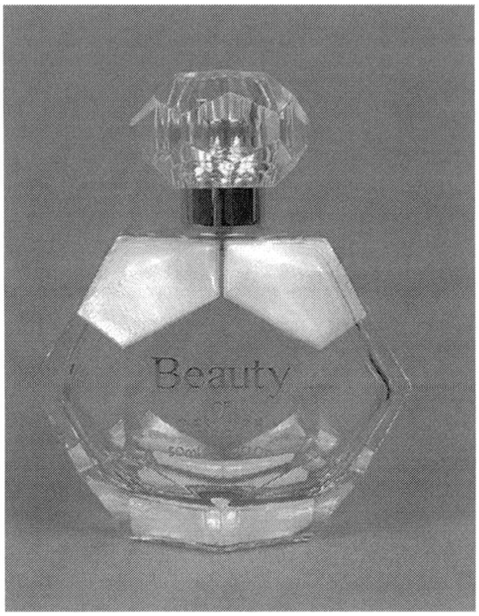

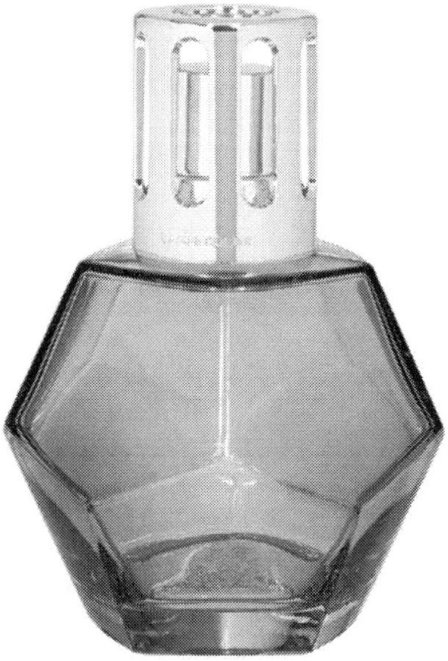

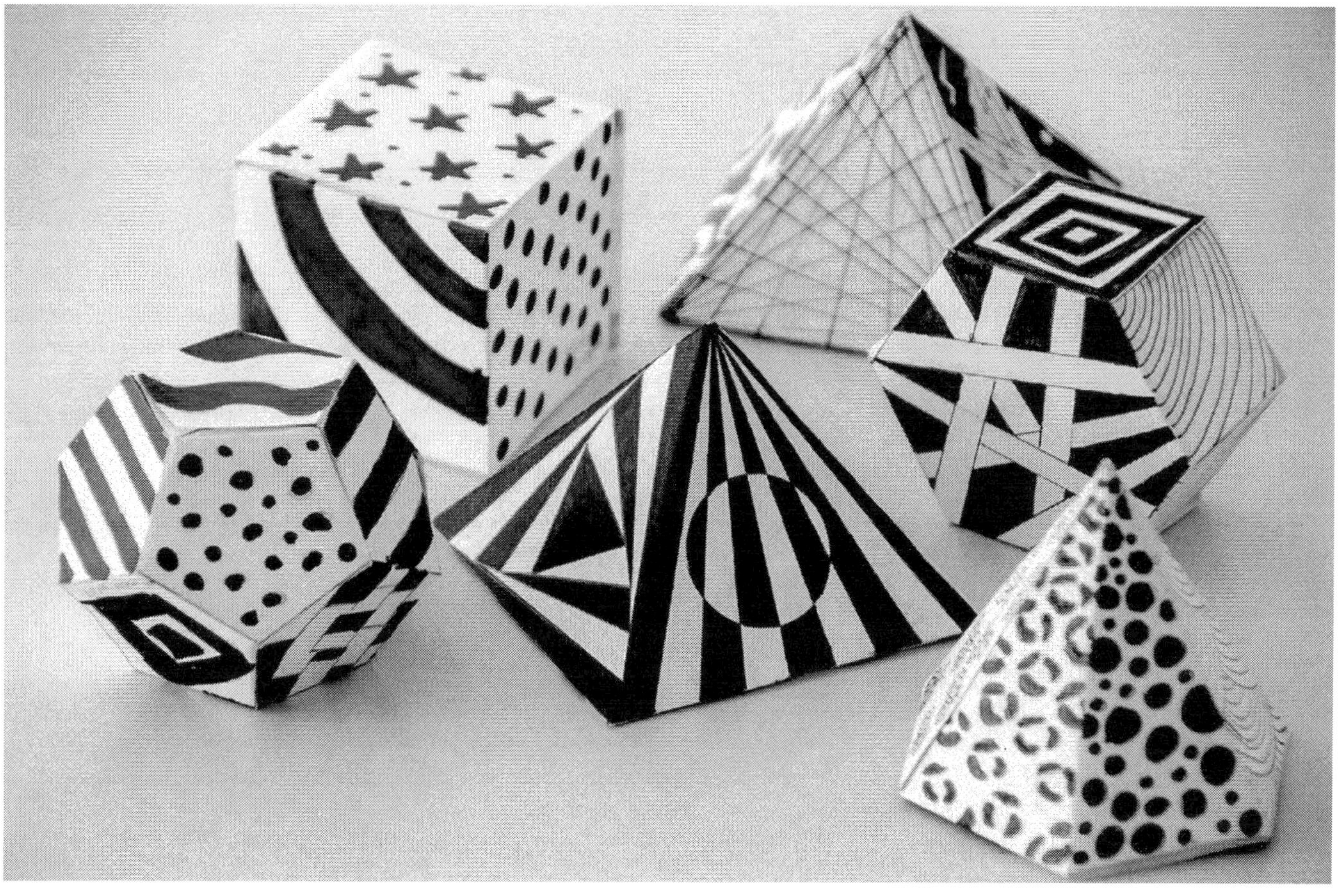

Elements & Principal Forms

Art vocabulary can be combined with forms. Each side of a form can be decorated to explore art elements like line, shape, color, texture, etc. Or to illustrate a variety of elements or principles.

Image Above Credit: www.arteascuola.com

Escher 3D Tessellation

The five platonic solids have all the same shape on each face. This is an opportunity to explore tessellations in 3D. Many forms contained here are made from triangles, squares, and parallelograms. These too can be tessellated. A video resource for creating tessellations can be found here: www.bit.ly/TessellationPlaylist

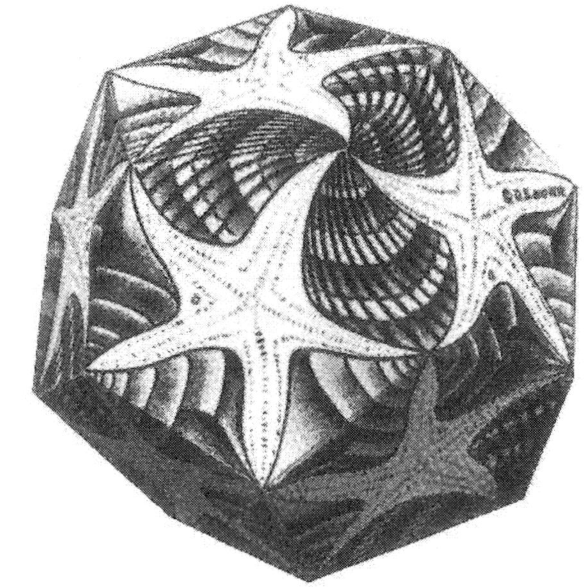

Stars and shells by M.C.Escher

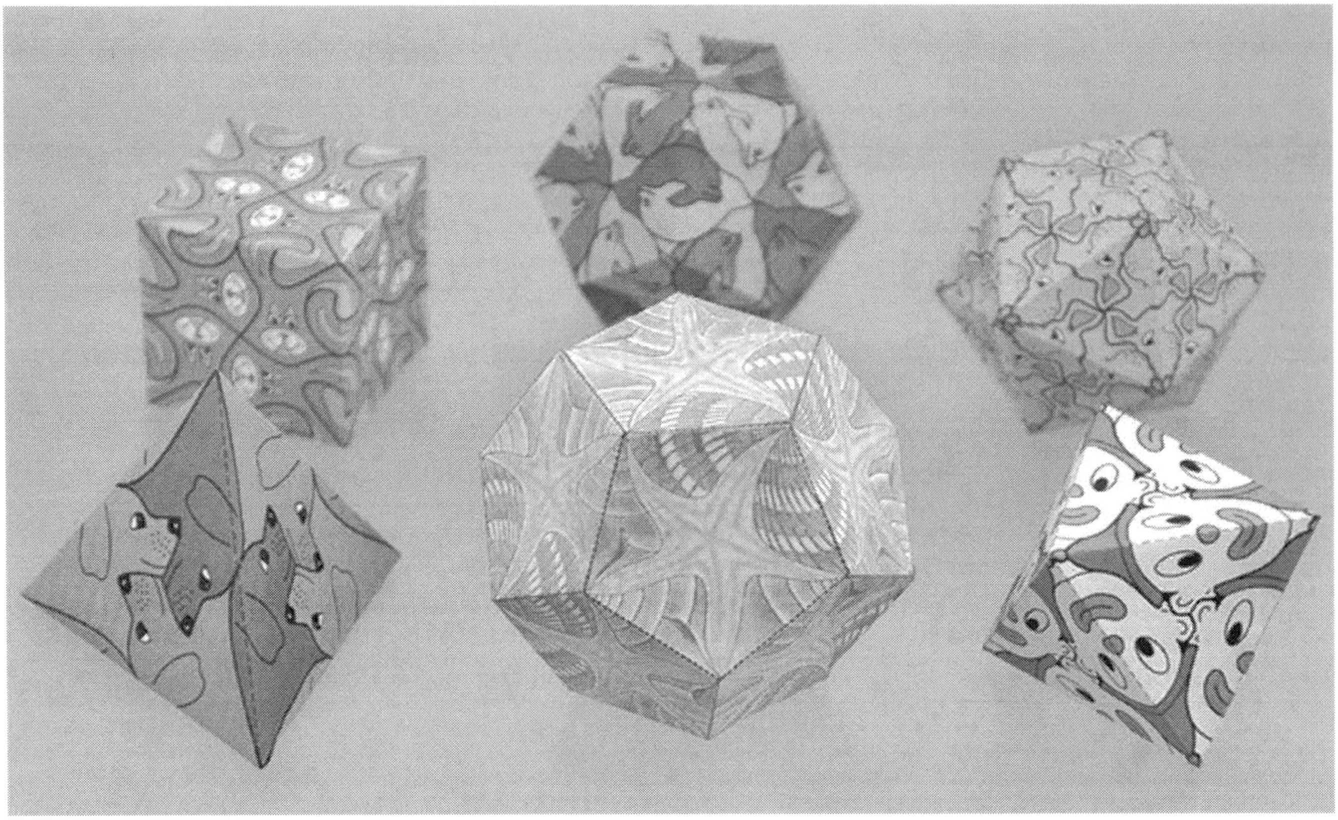

Image Credit www.polyhedr.com

Stuffed Felt Forms

Sewn pillows can be made of felt or other fabrics filled with stuffing. The sample image here was done with a blanket stitch by Ann Rinkenberger author of harvestmoonbyhand.blogspot.com.

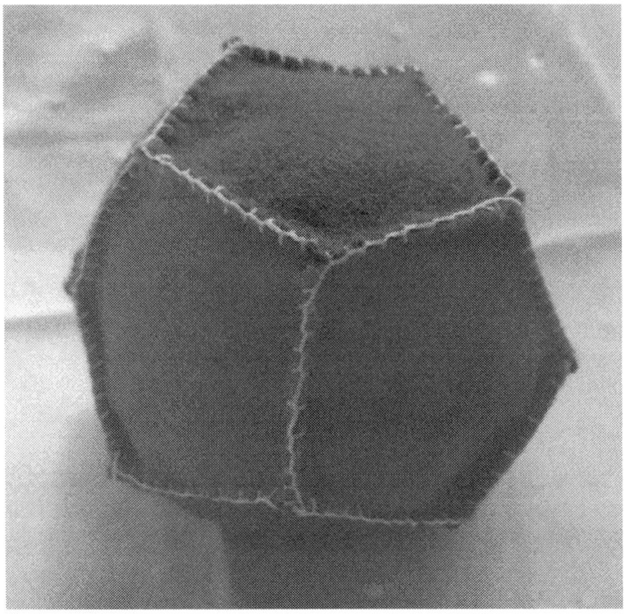

By Ann Rinkenberger author of
harvestmoonbyhand.blogspot.com/
http://harvestmoonbyhand.blogspot.com/2013/06
/hand-embroidered-geometry-felt.html

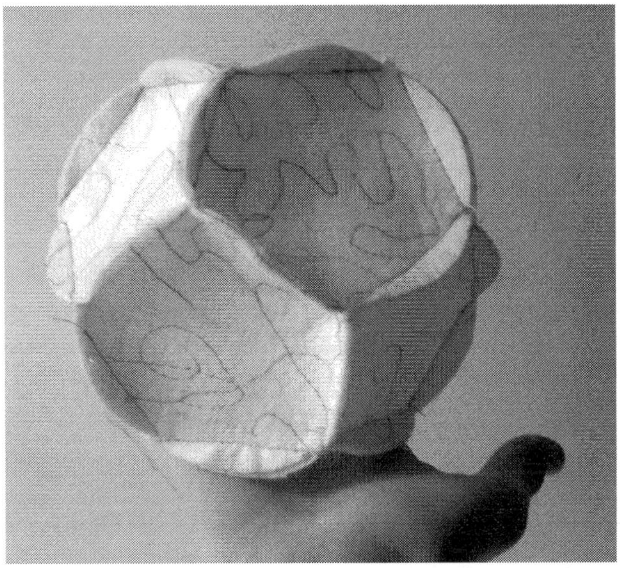

www.instructables.com/Sew-a-Dodecahedron/

Lamp Forms:

Battery operated LED lights can be placed inside forms with pierced surfaces or made with stained glass. Those these samples are traditional Moroccan lamps, they could be made with clay, wood, or other rigid materials. Use caution when using lights that can create heat or be plugged in.

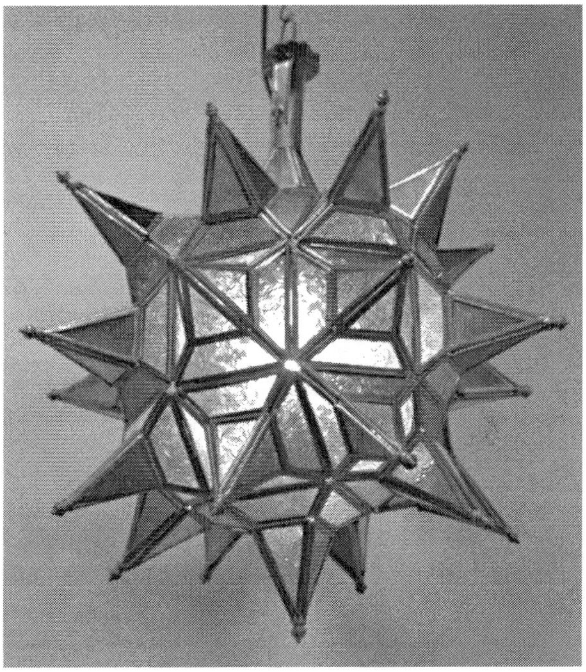

Forms In Relief

Thick clay or plaster forms can be carved in relief to create expressive works of art. Each side of the form could represent a chapter in the artist's life, positive or negative. The example below is by Mark Ryden.

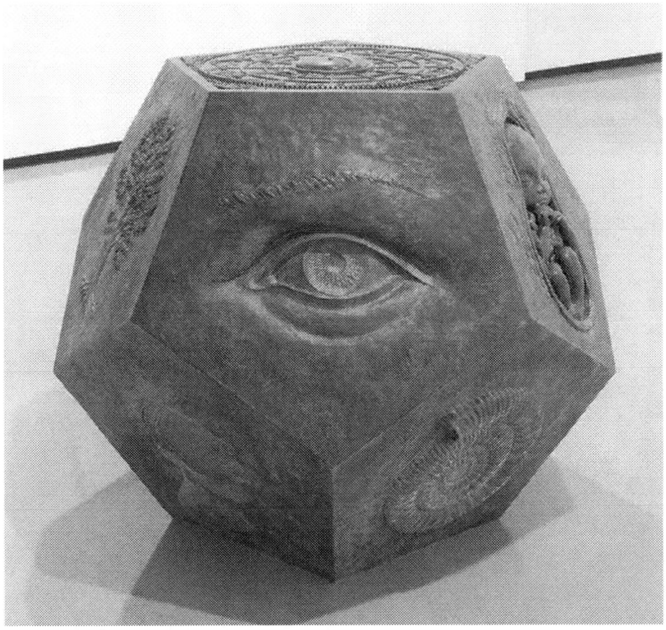

Mark Ryden is a contemporary artist that uses form for the basis of sculpture. Artsy.com image

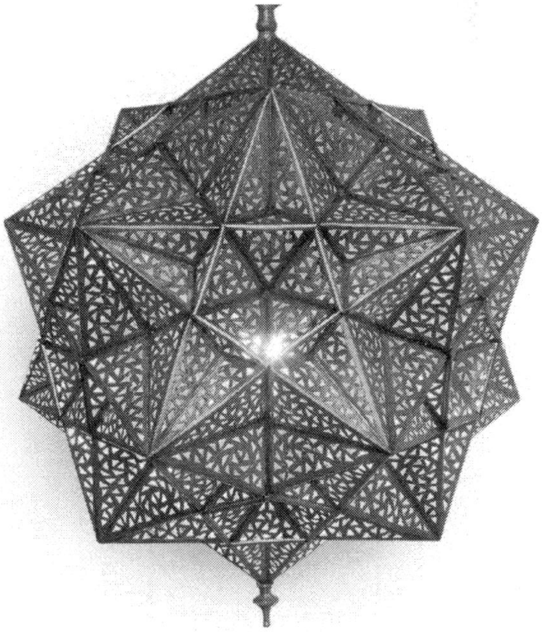

Notes/Ideas:

Cone

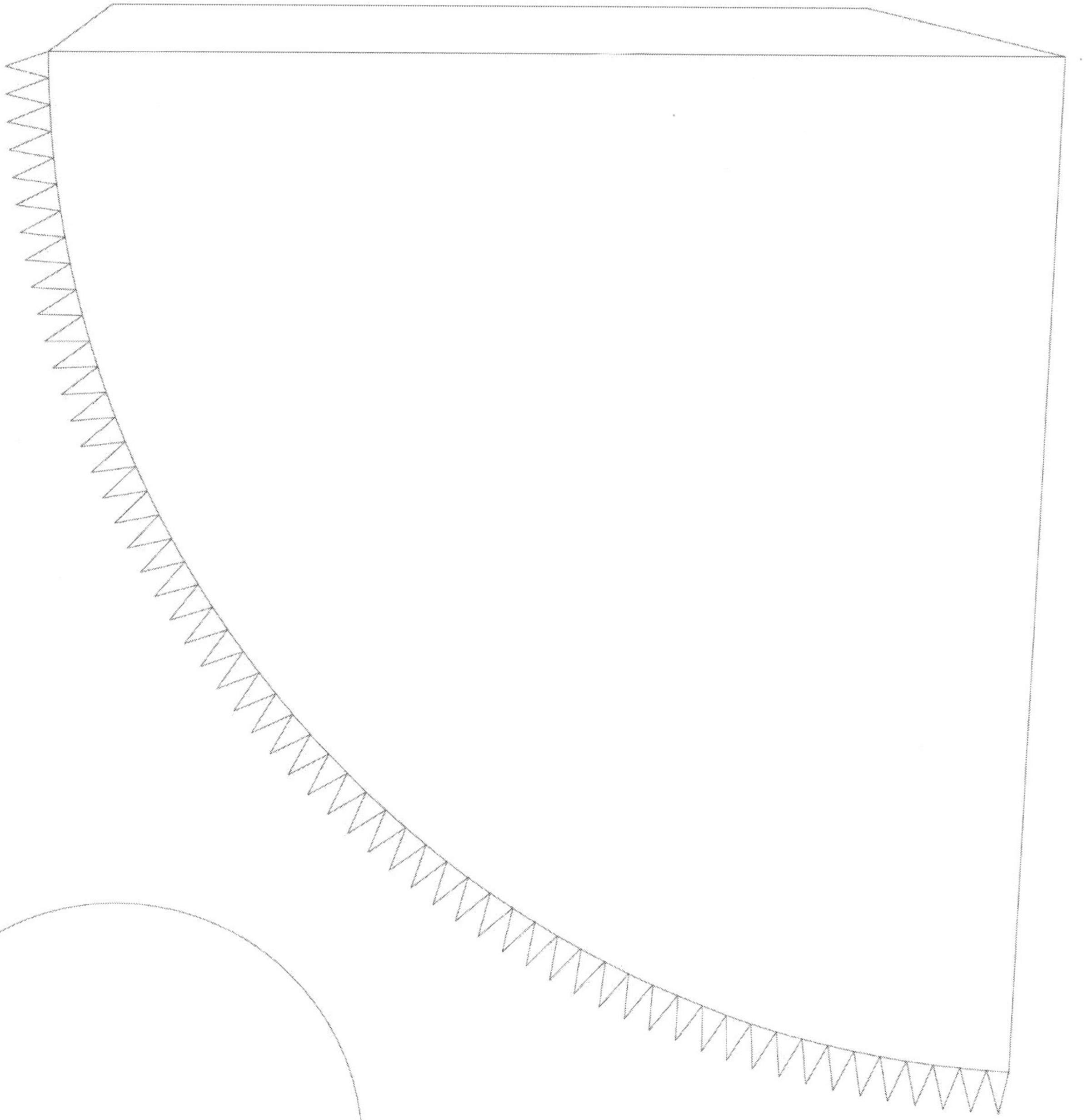

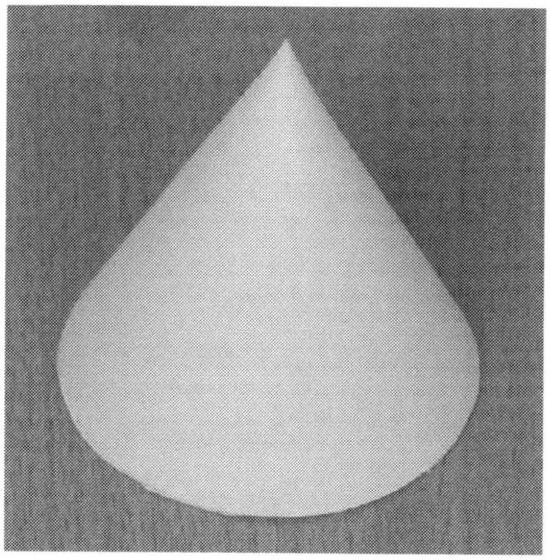

Triangular Prism Box

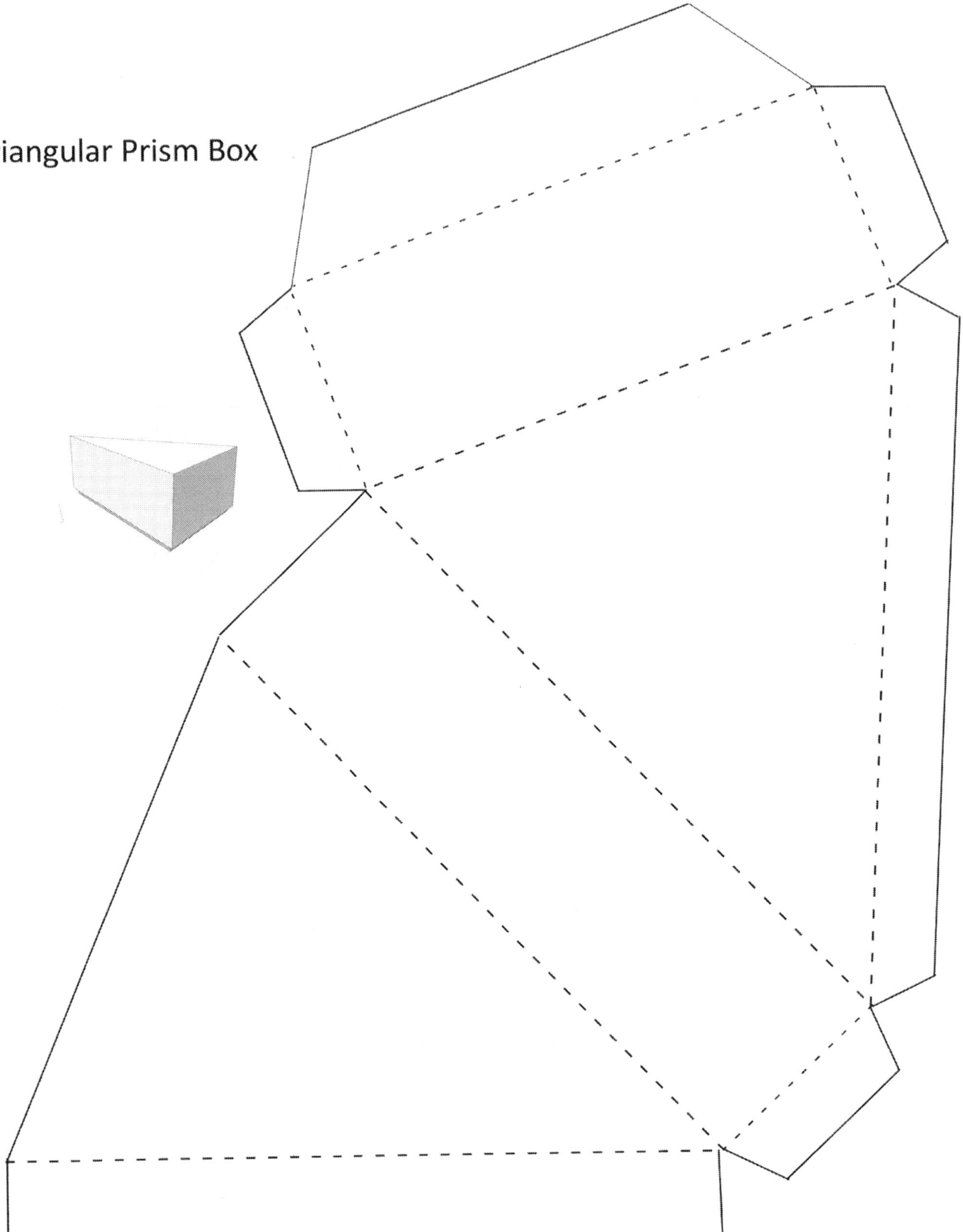

Cube

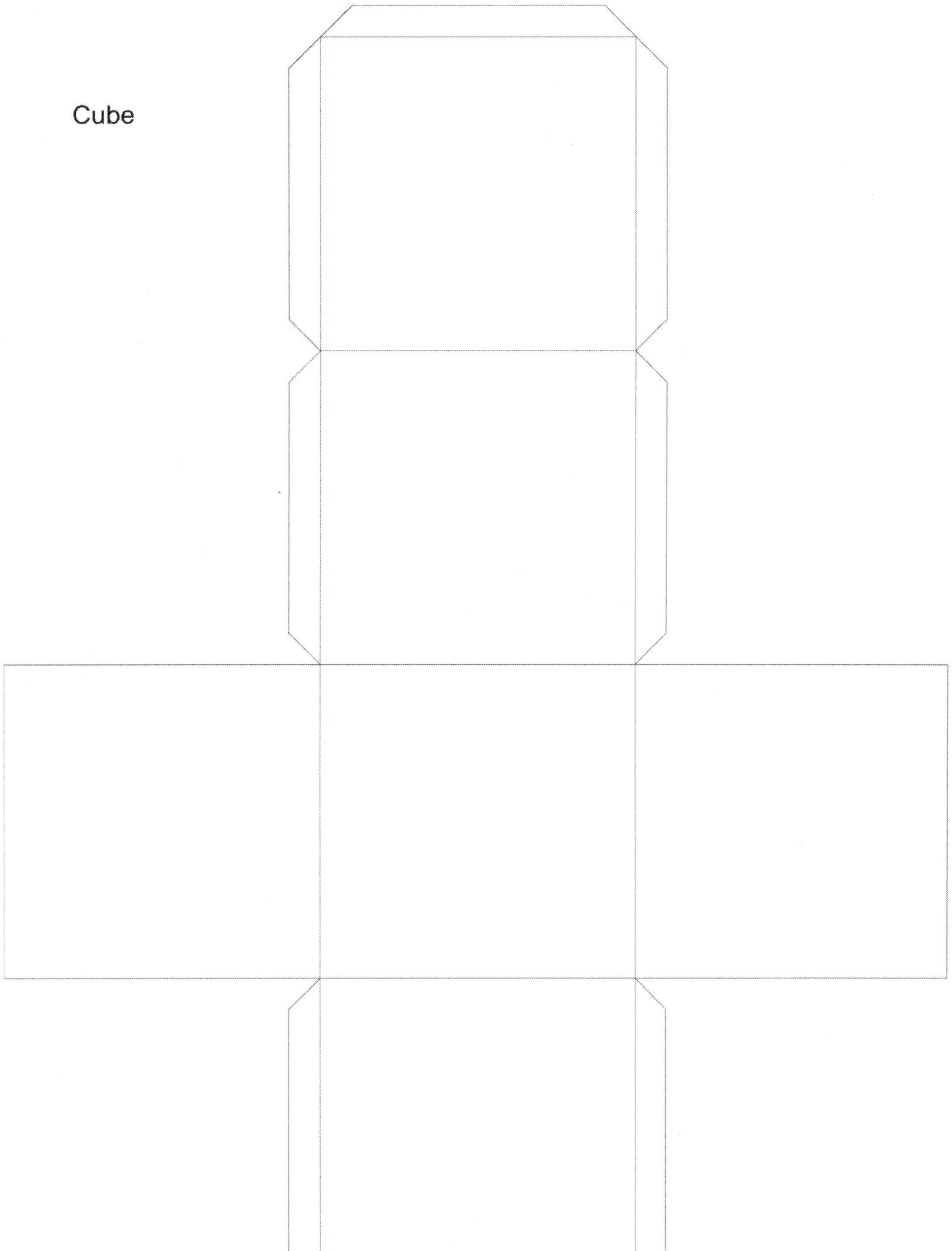

Cylinder

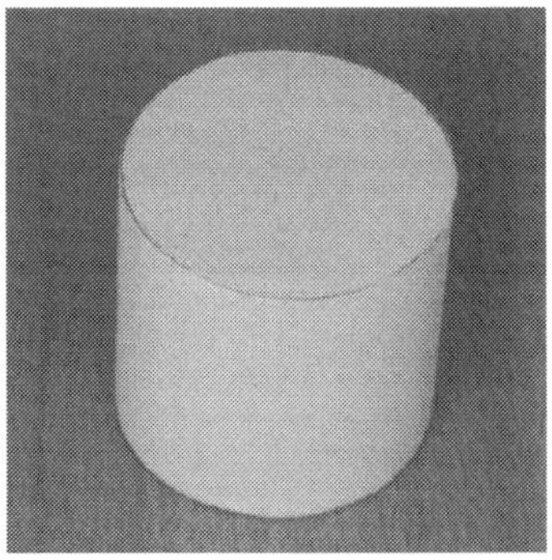

Square Dipyramid

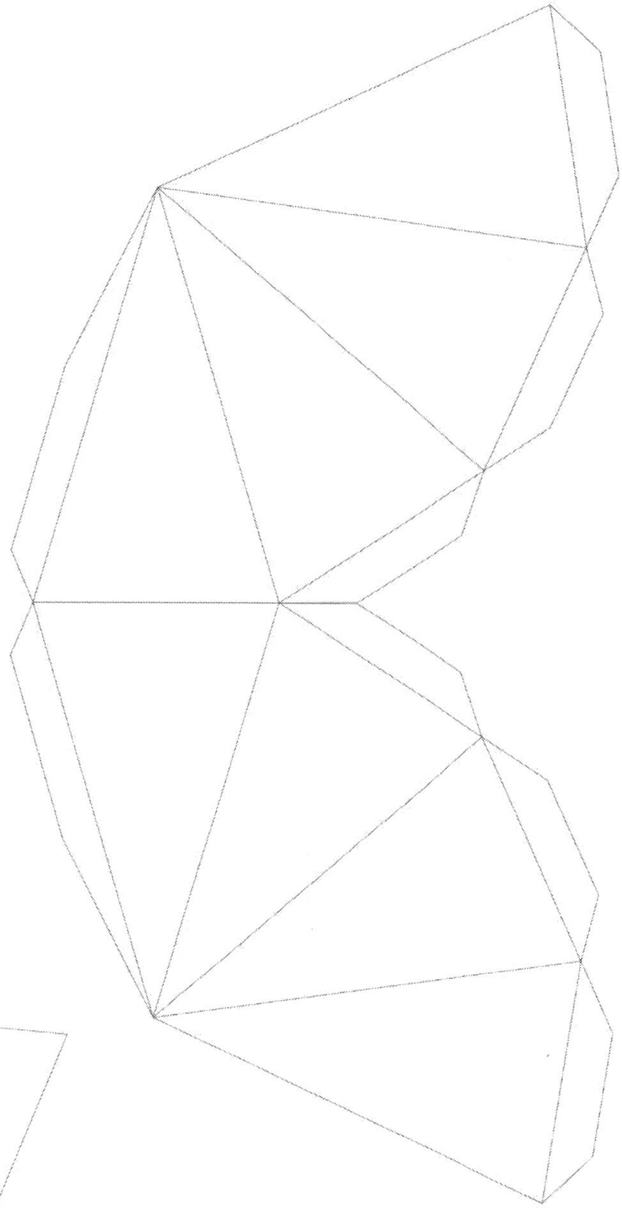

Triangular
Dipyramid

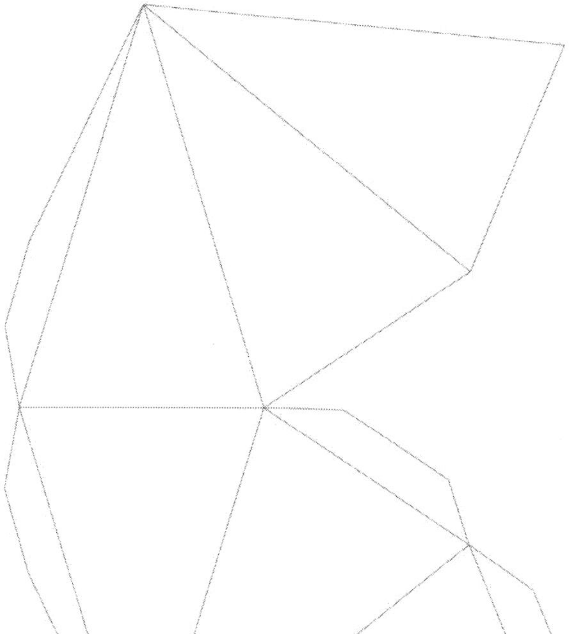

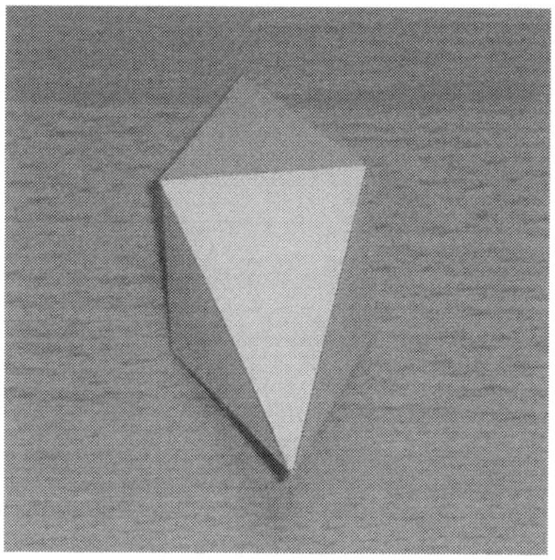

Oblique Rectangular Prism

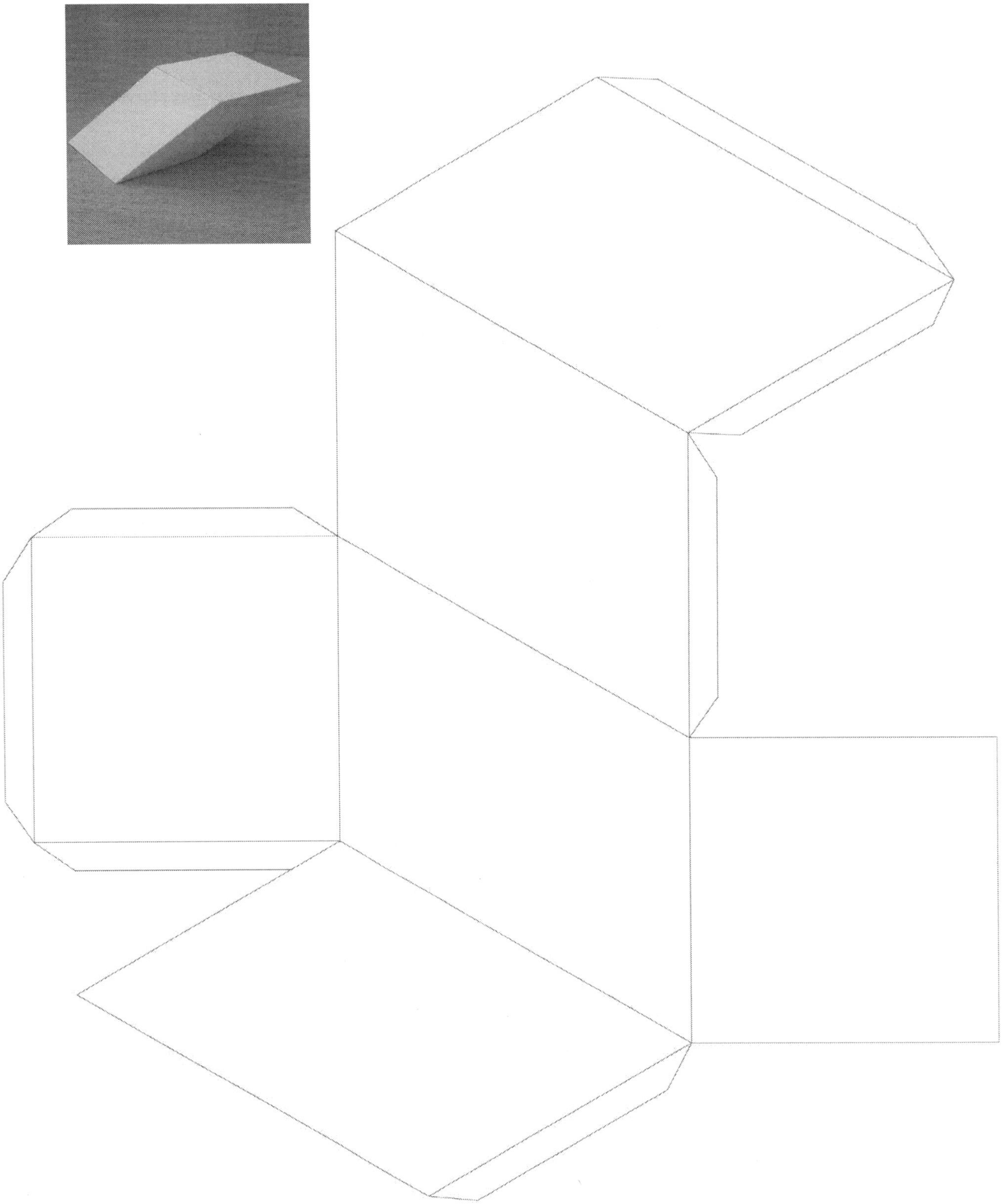

Oblique Cylinder

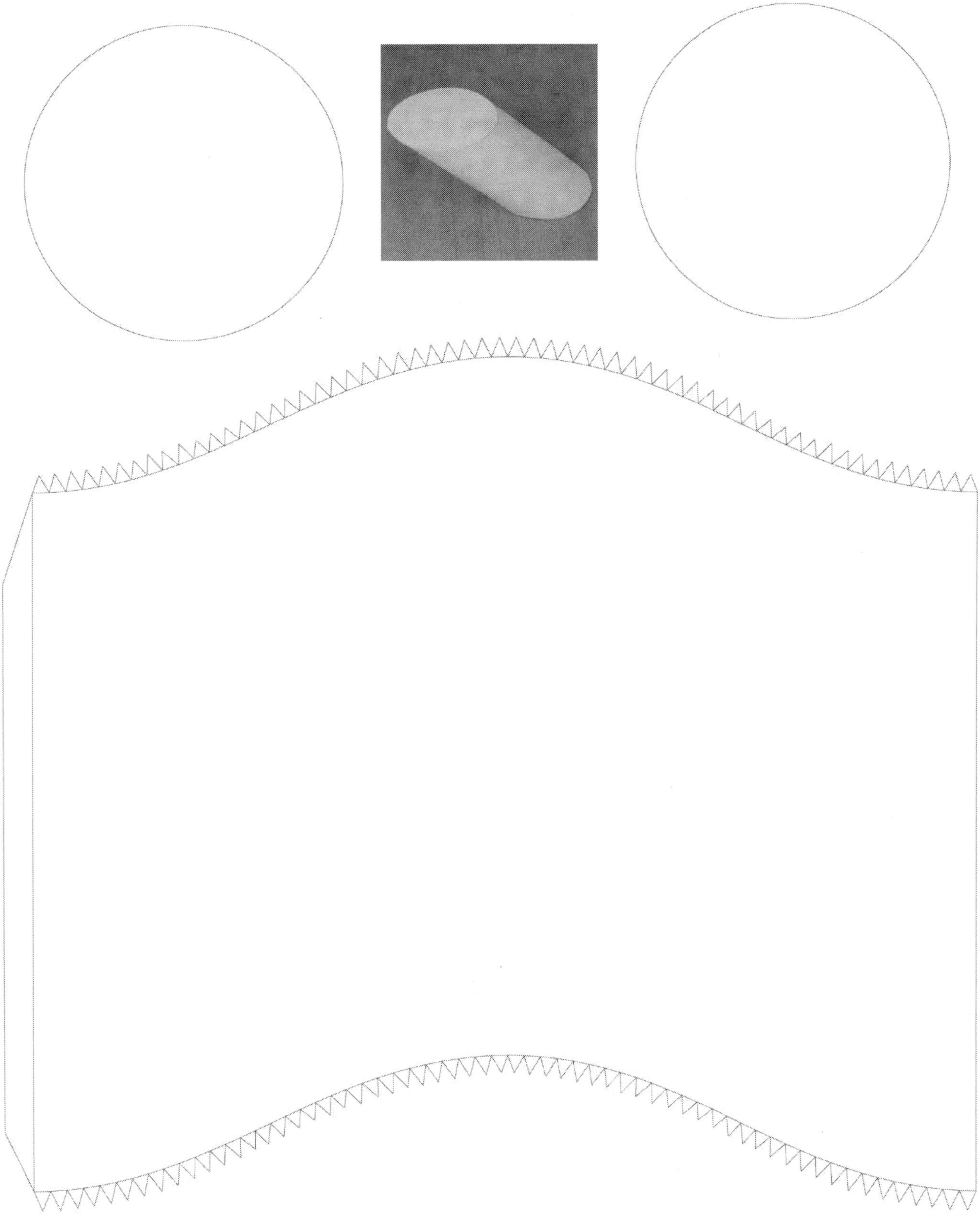

Octahedron

Pentagonal Dipyramid

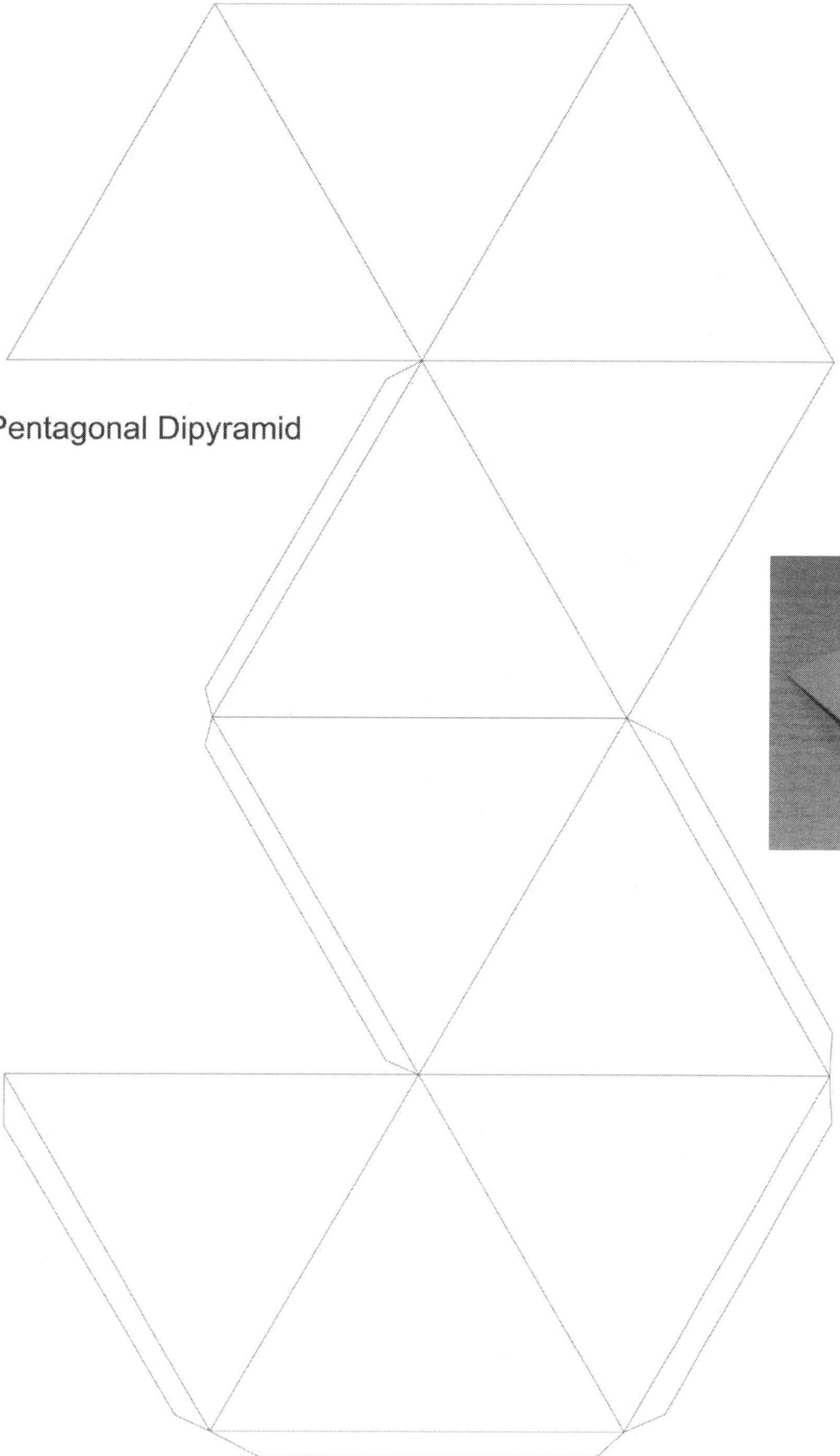

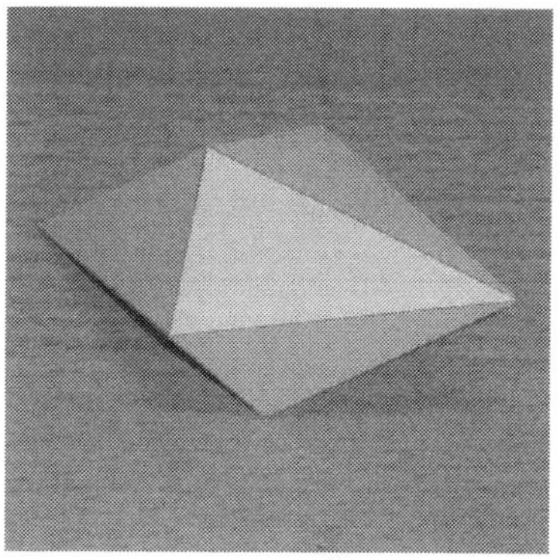

Pentagonal pyramid

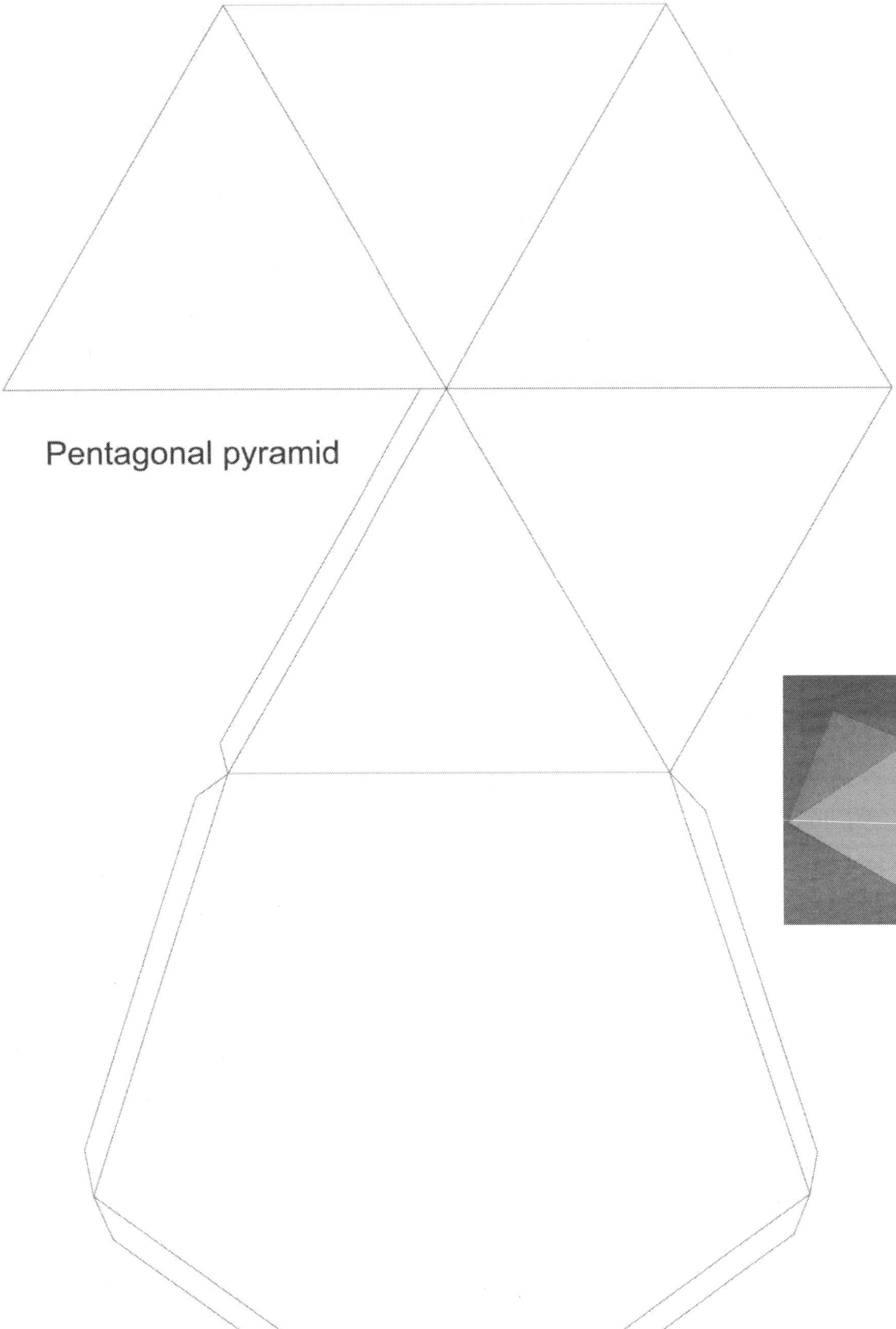

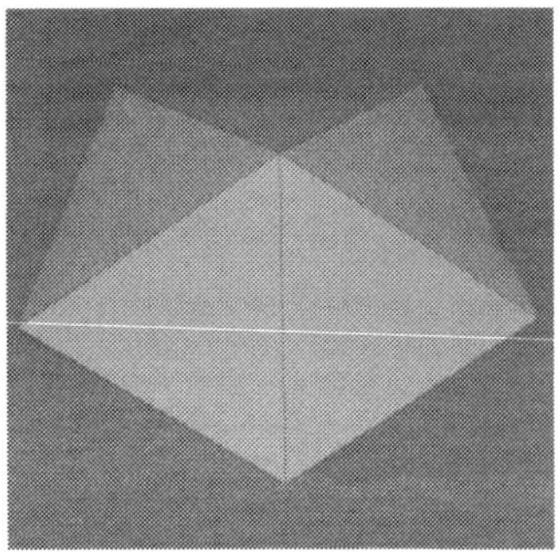

Pyramid

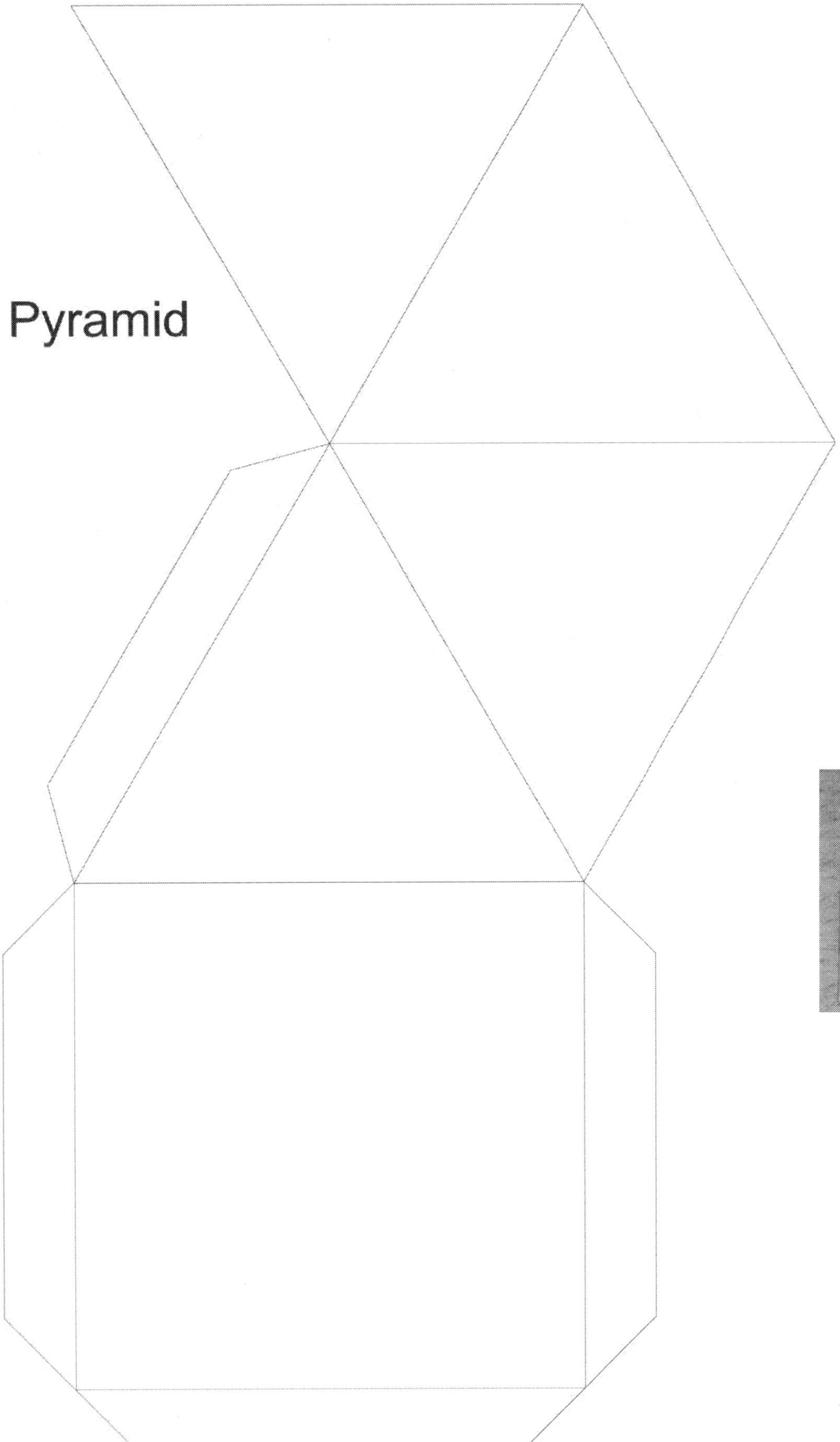

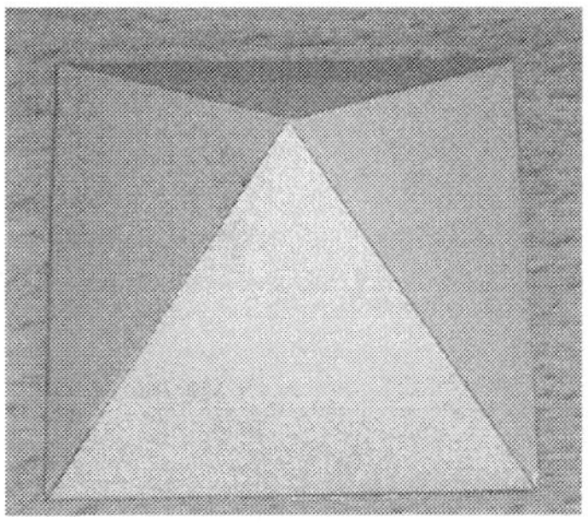

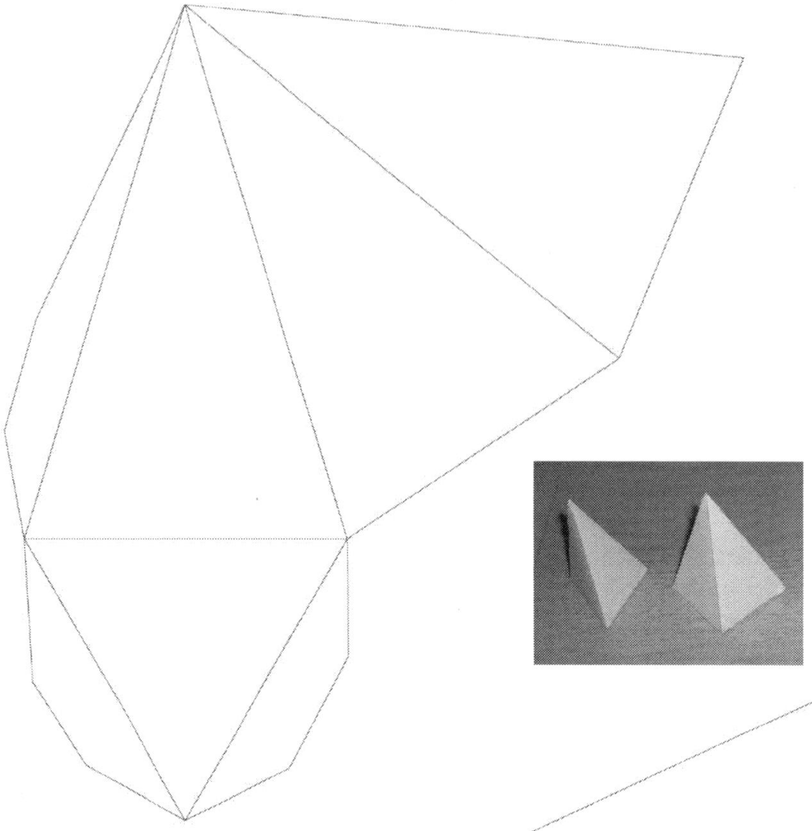

Triangular
Pyramid

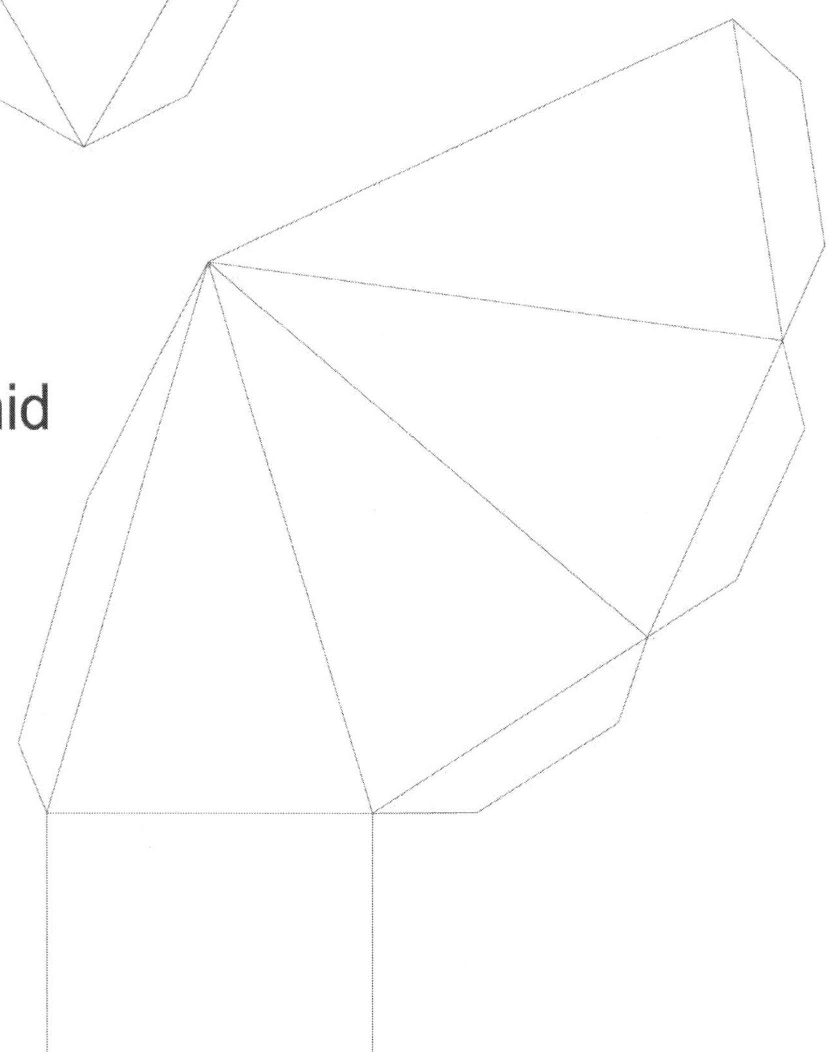

Square Pyramid

Rectangular Prism

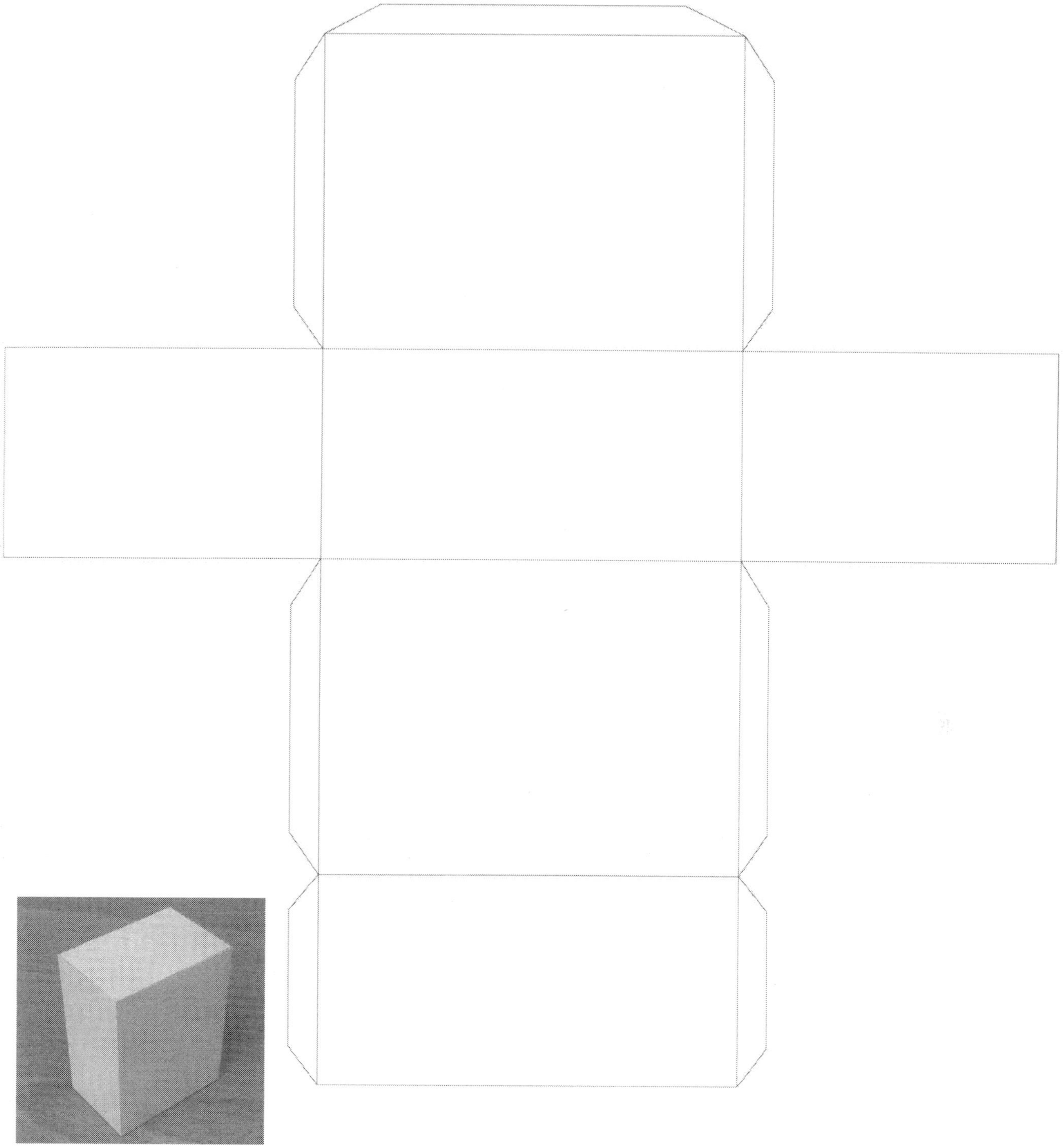

Sphericon

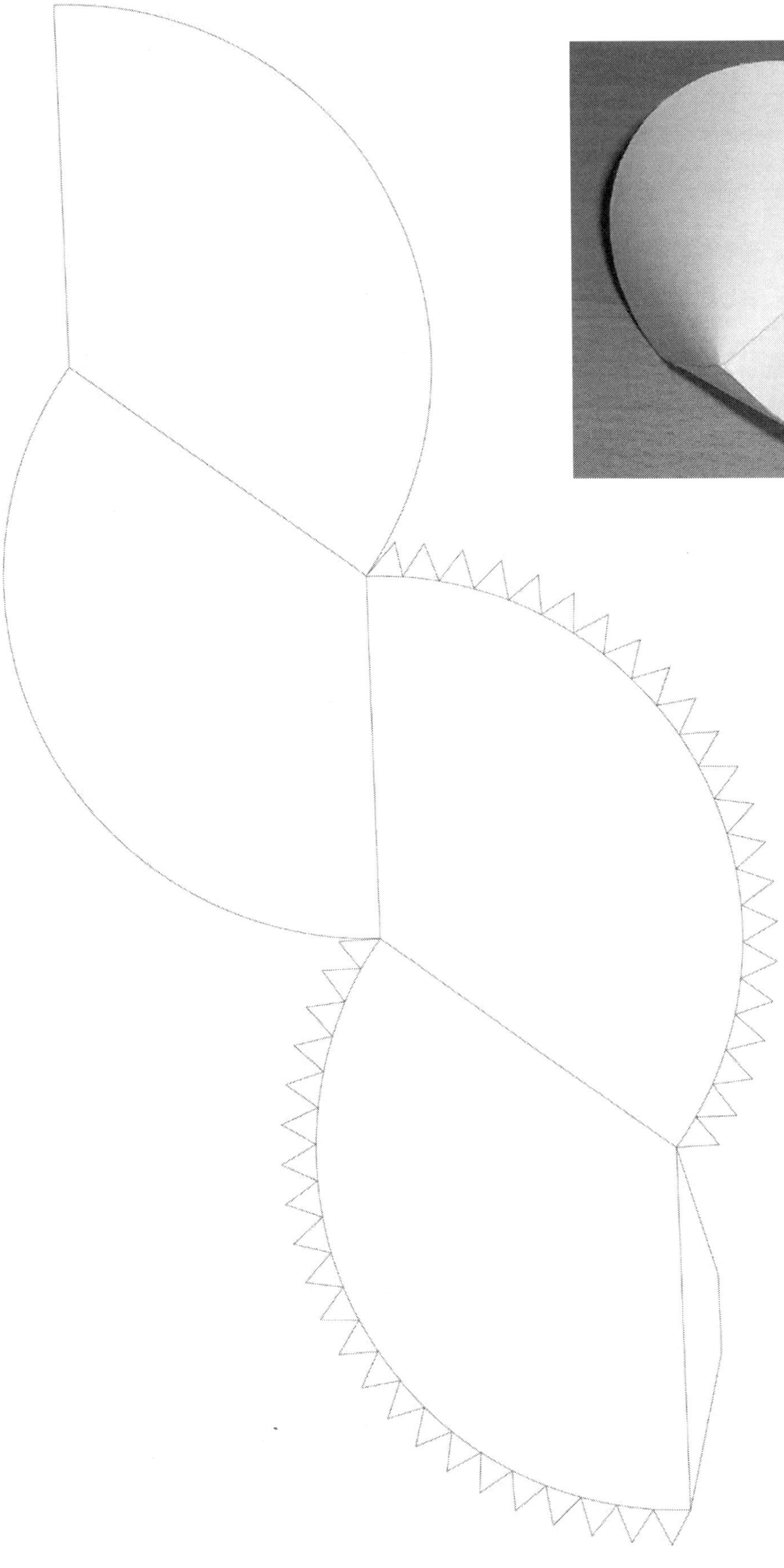

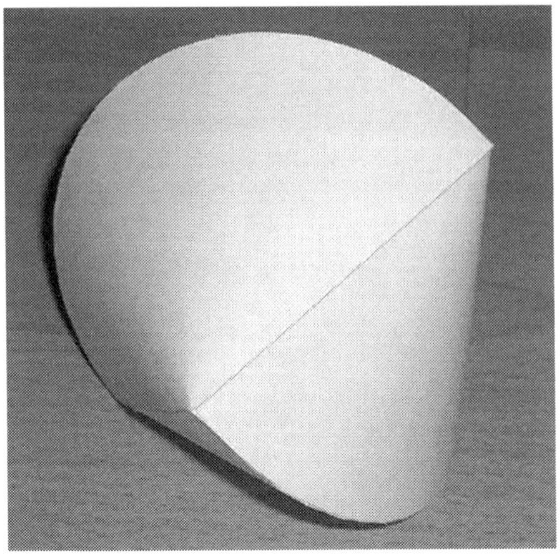

Square Antiprism

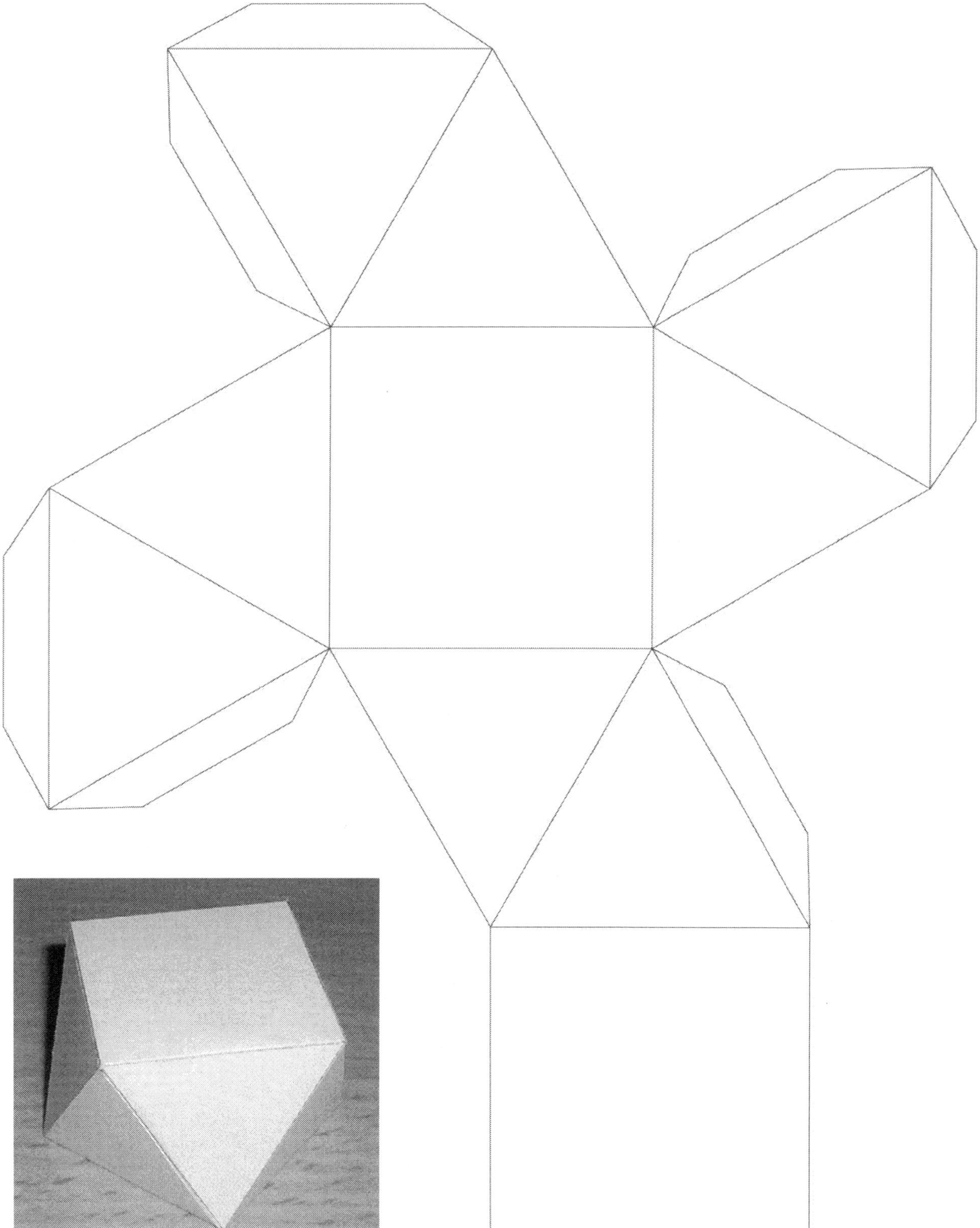

Tetragonal Trapezohedron

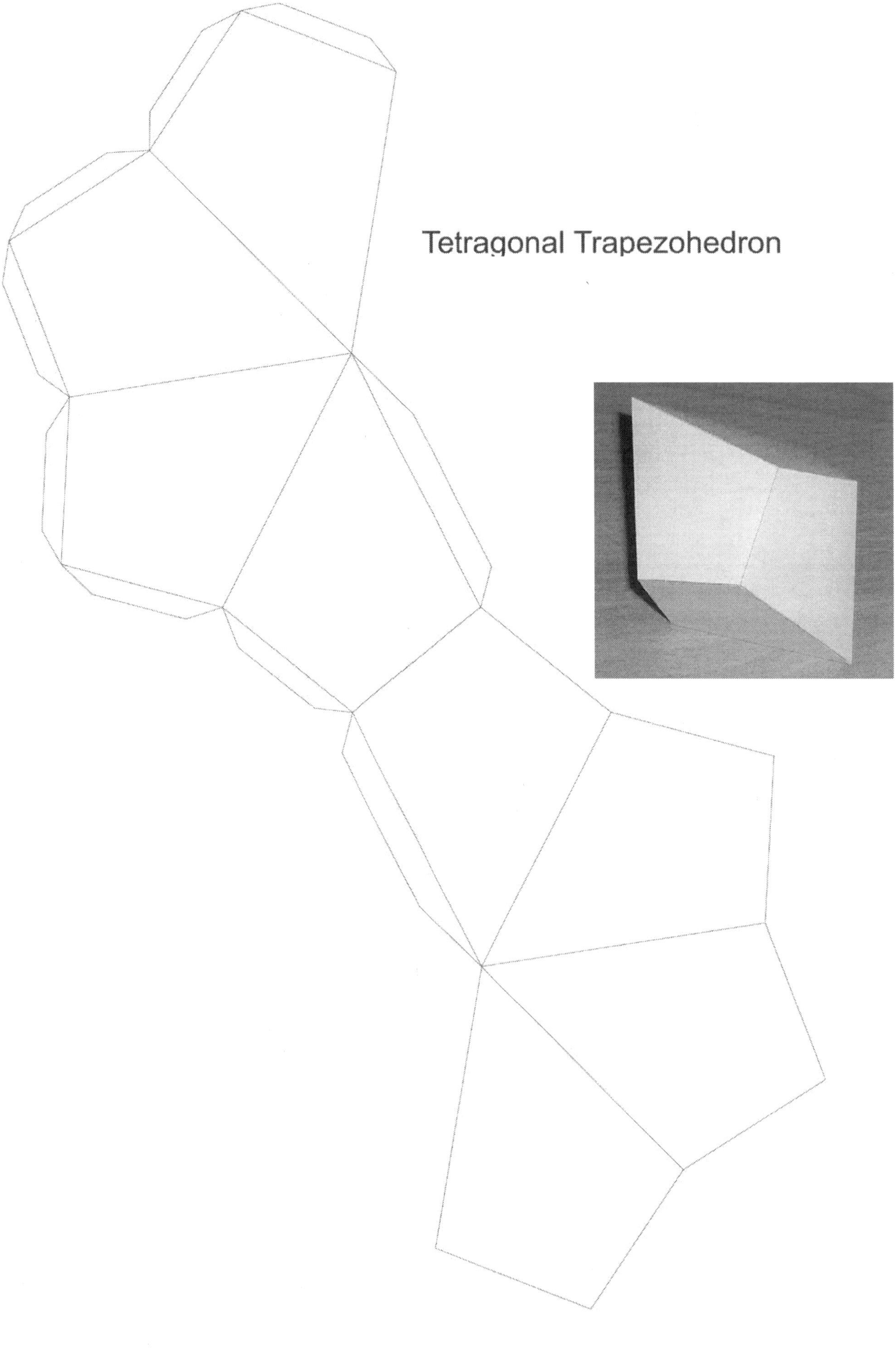

Tetrahedron

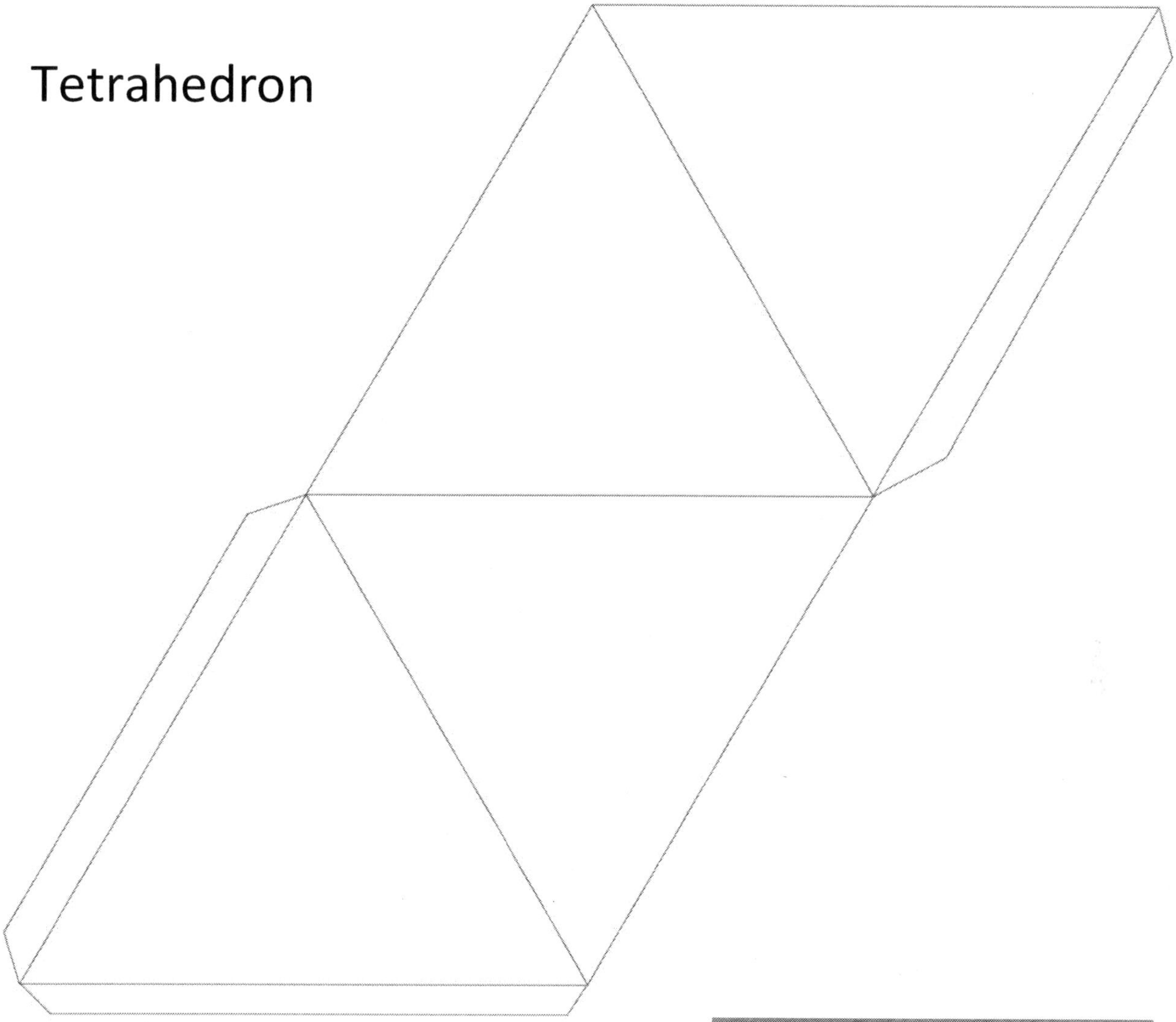

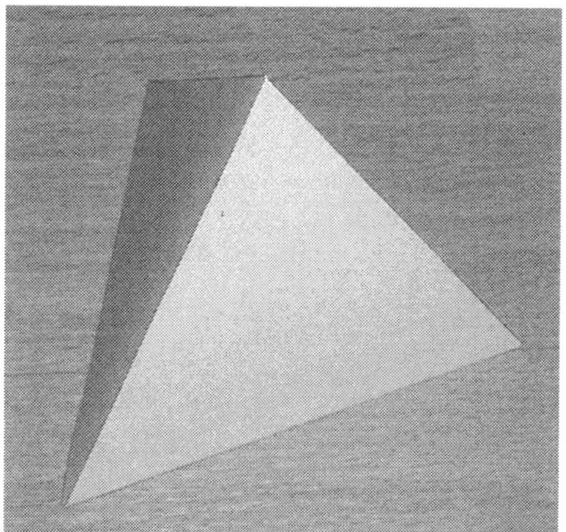

Diamond Form

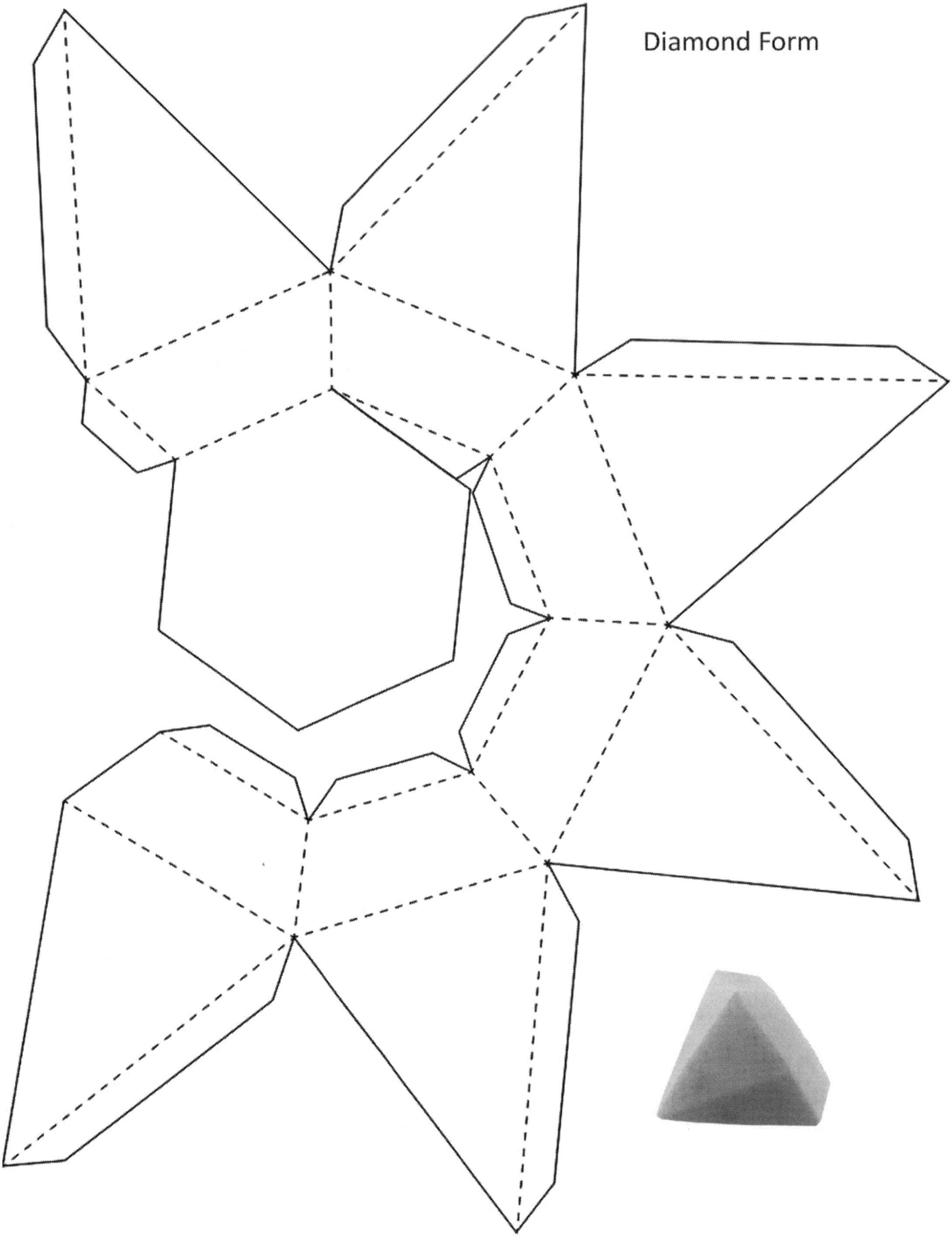

Trapezoidal Prism

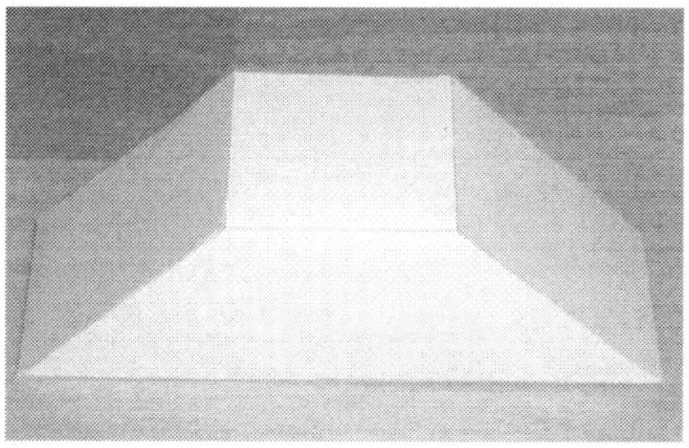

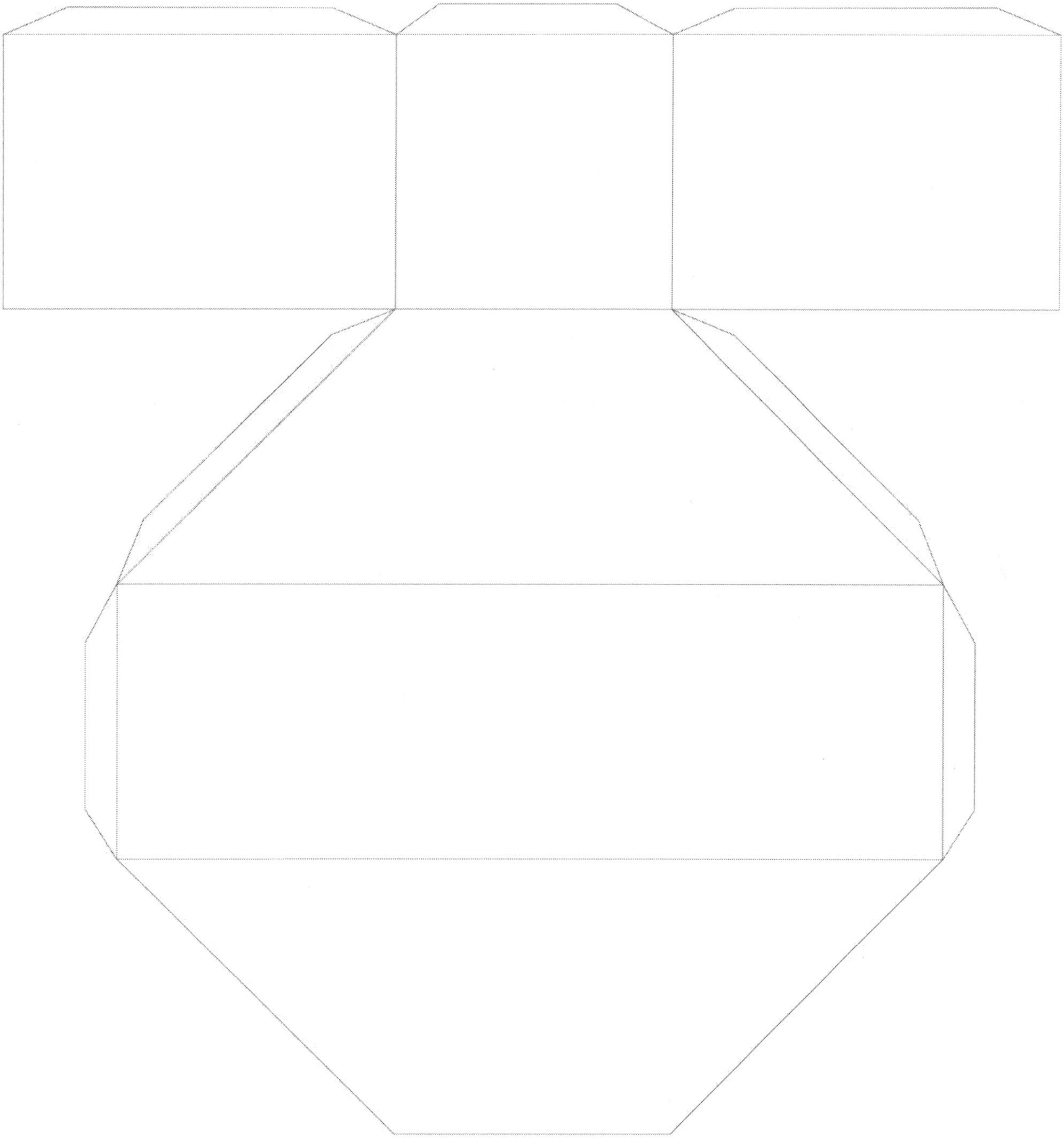

Triangular prisms

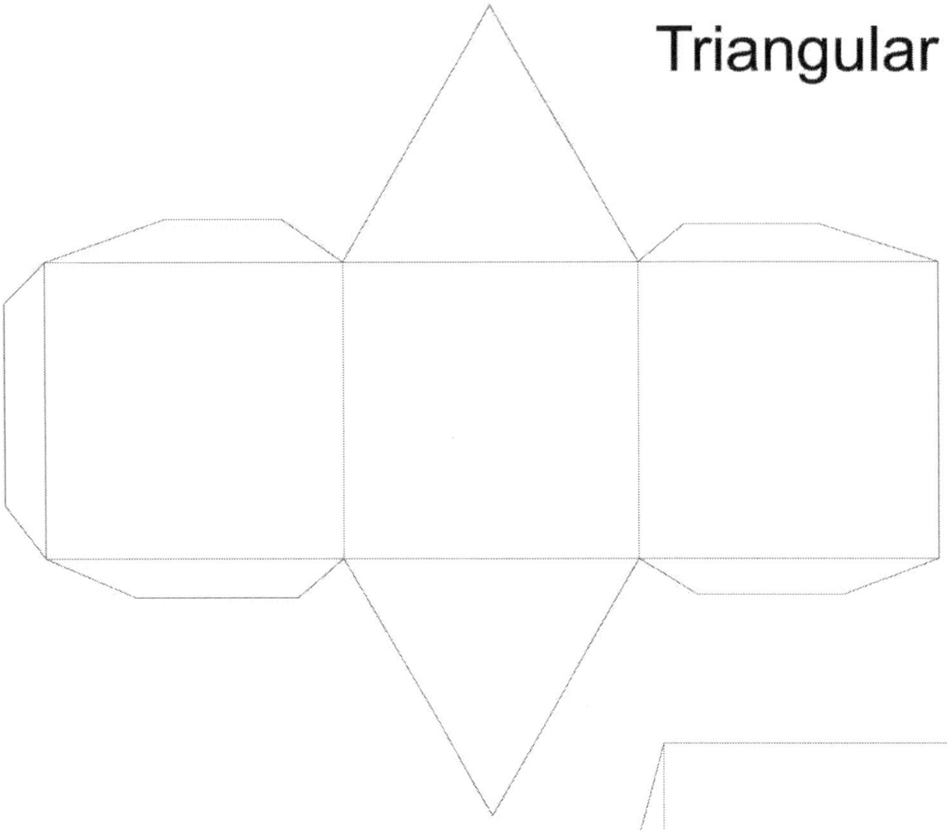

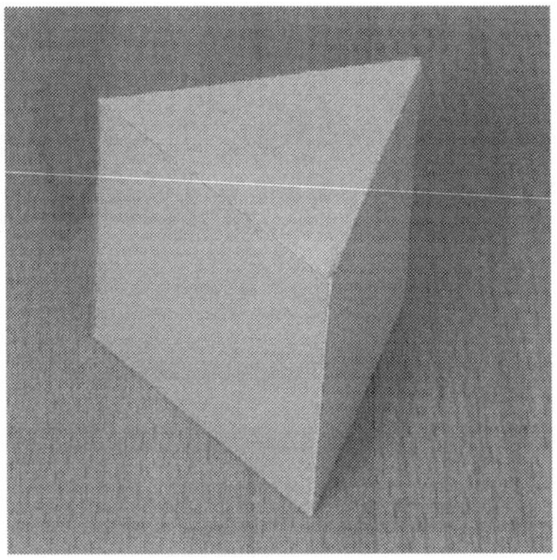

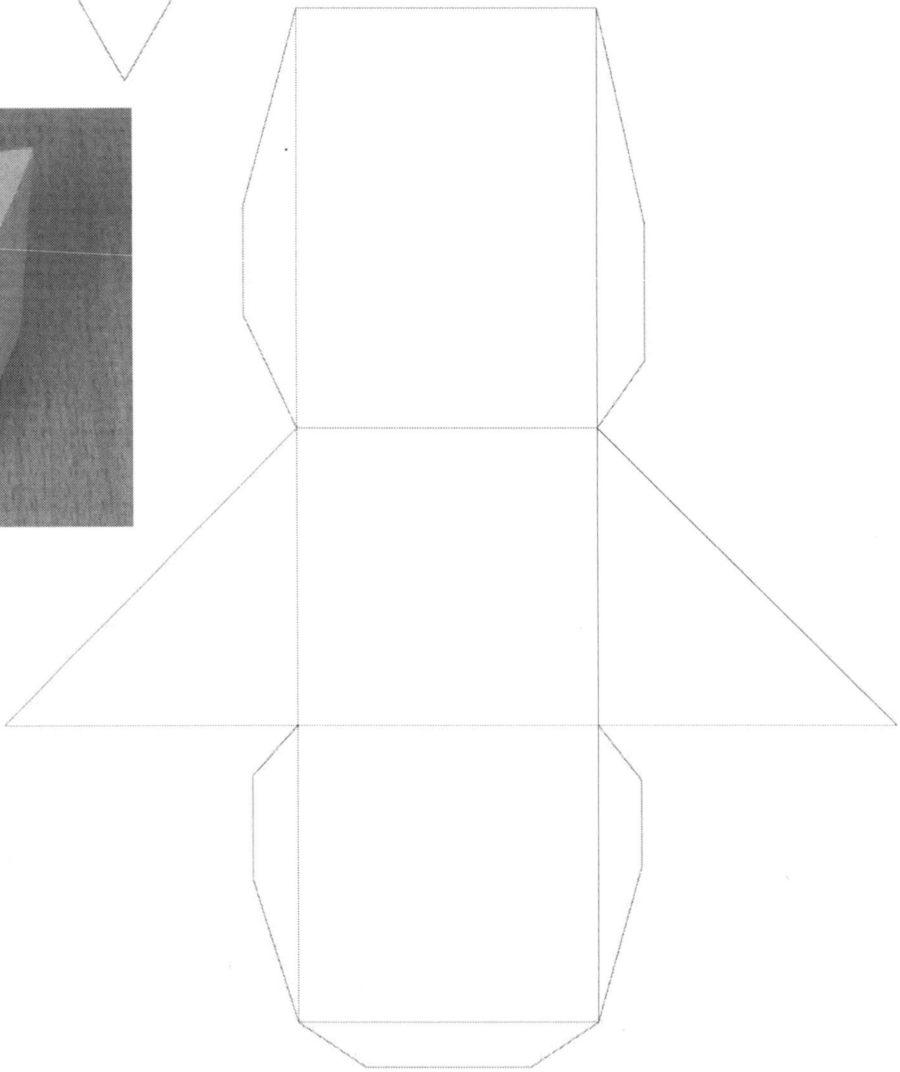

Truncated Tetrahedron

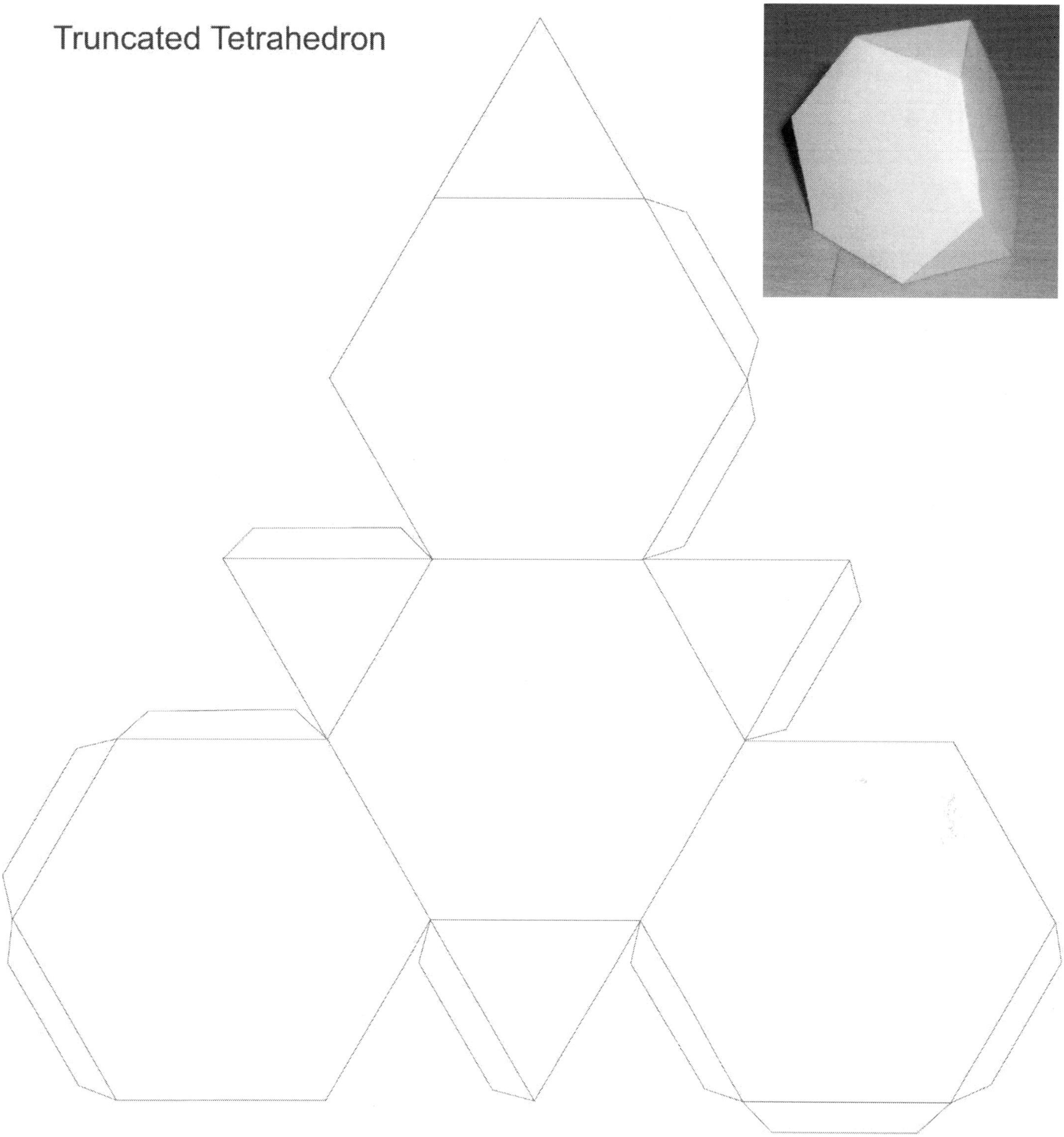

Twisted rectangular prism

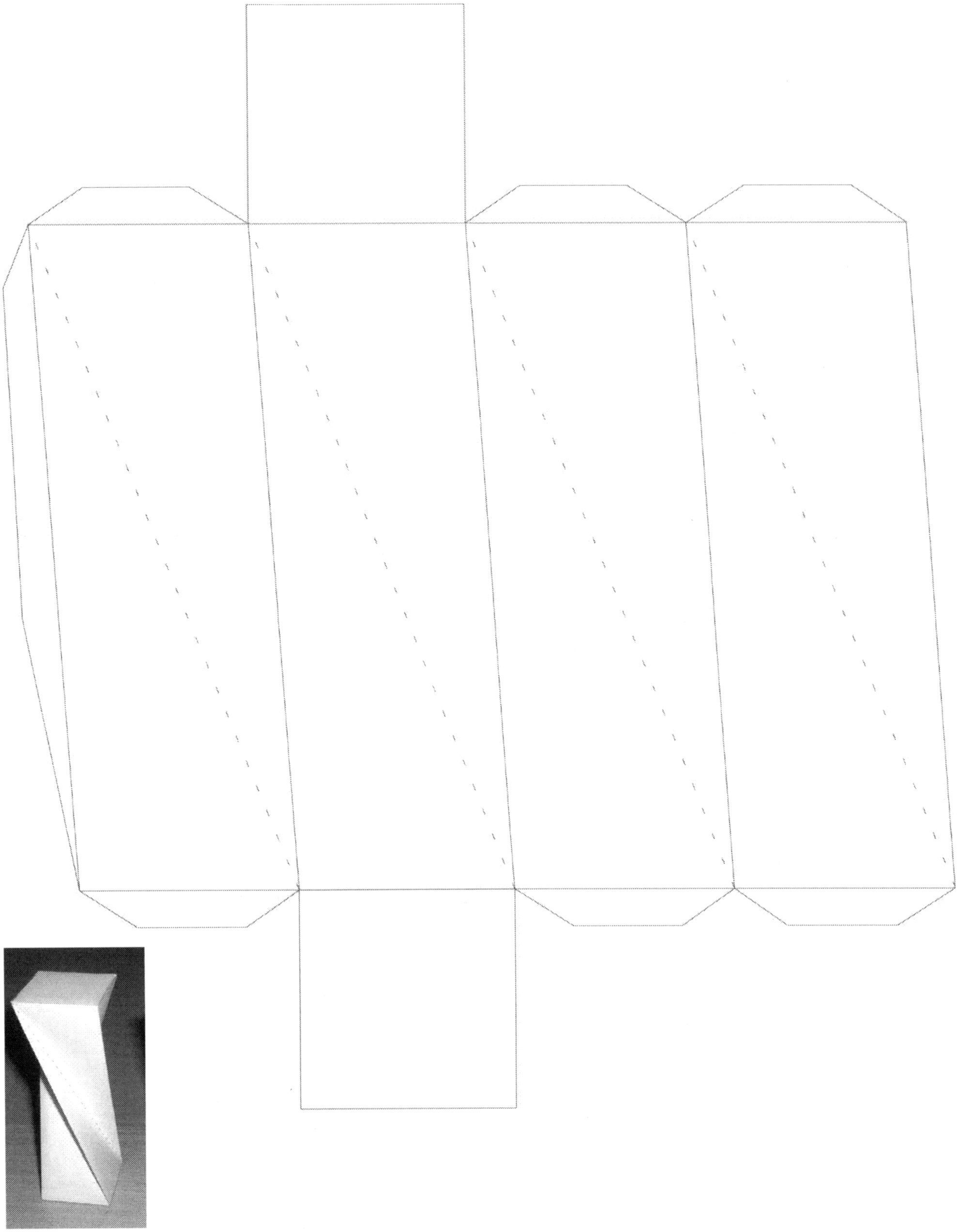

Compound Dodecahedron and Icosahedron

Fold the dotted lines forwards
Fold the other lines backwards

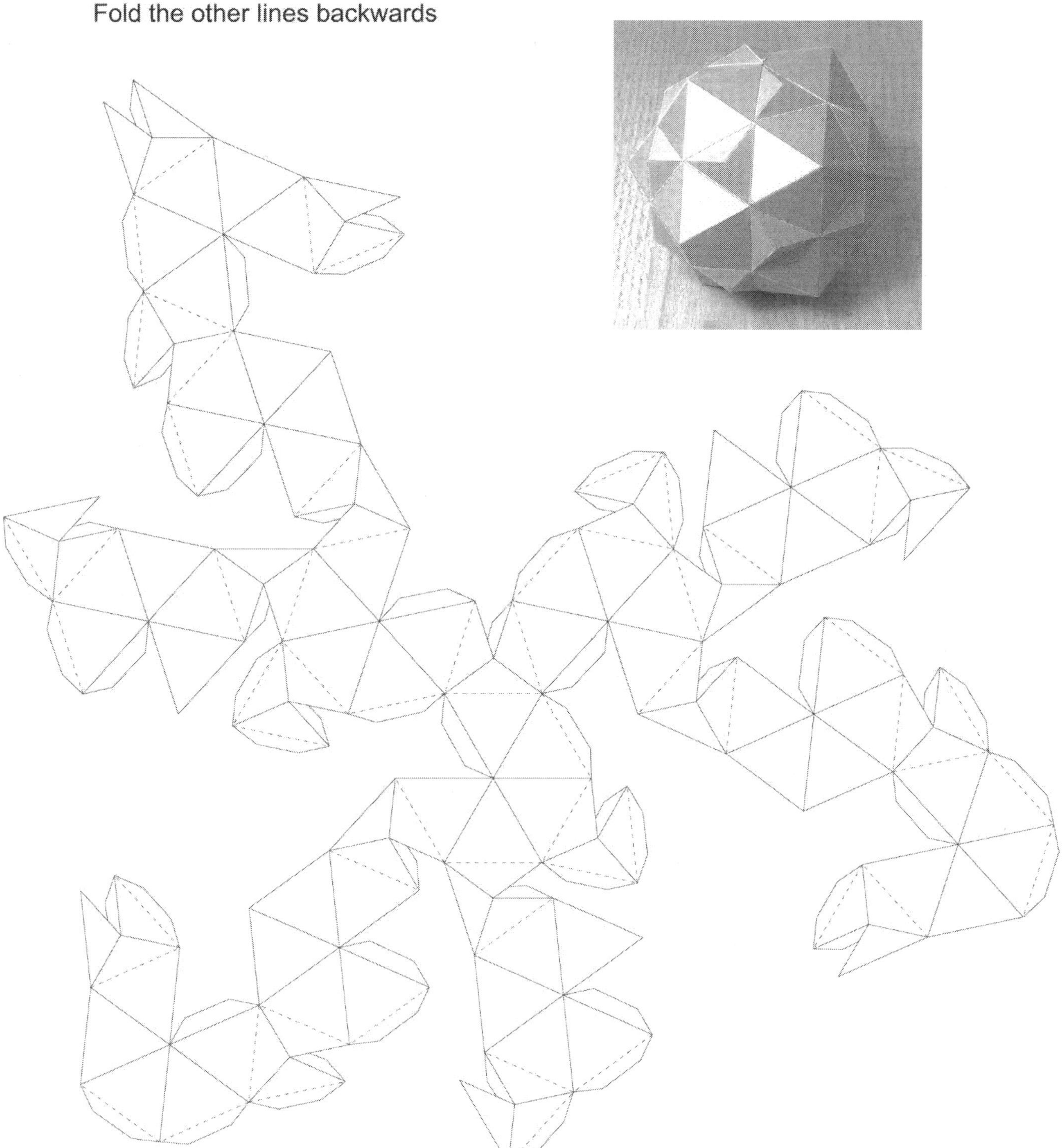

Compound of Cube and Octahedron

Fold the dotted lines forwards. Fold the other lines backwards

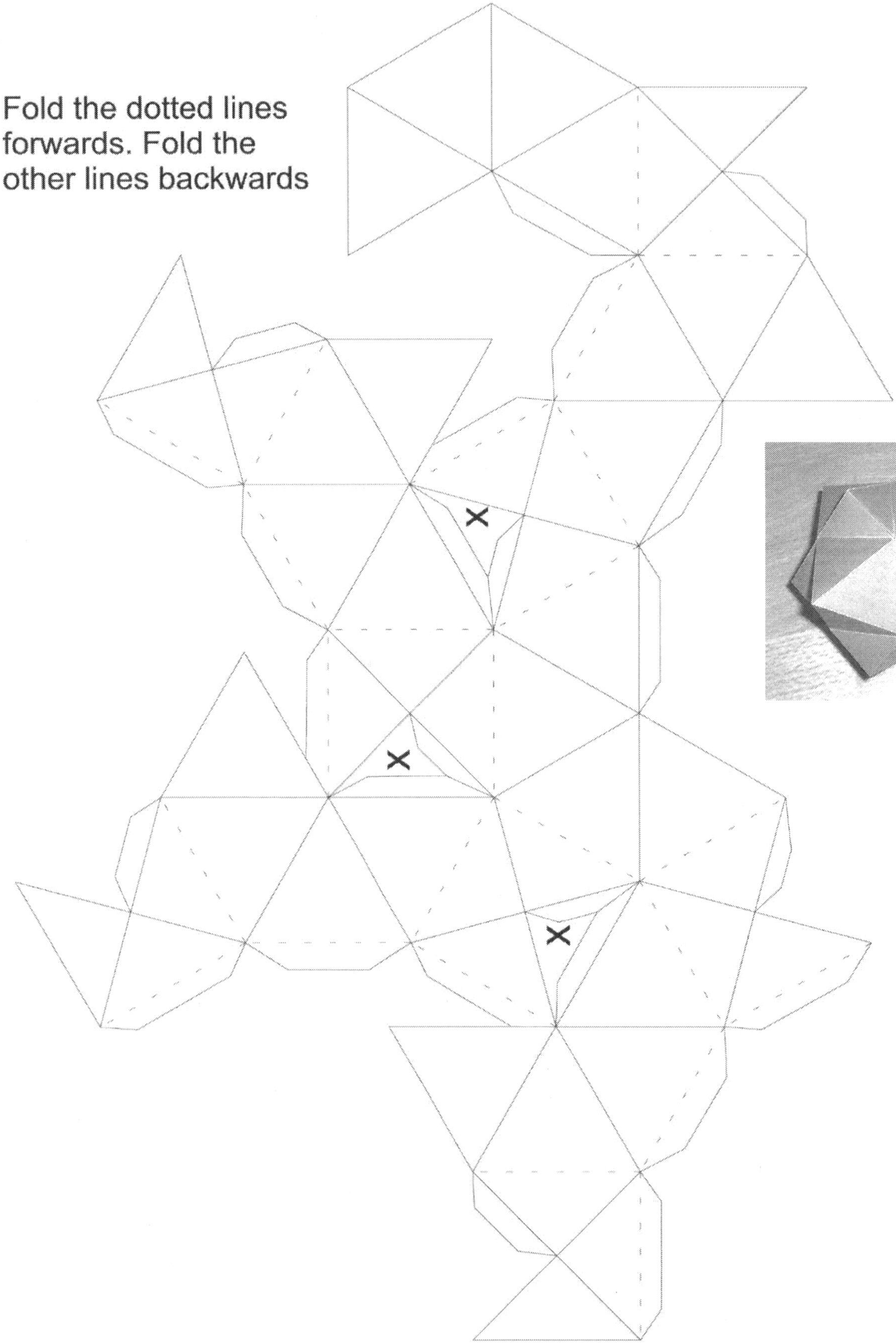

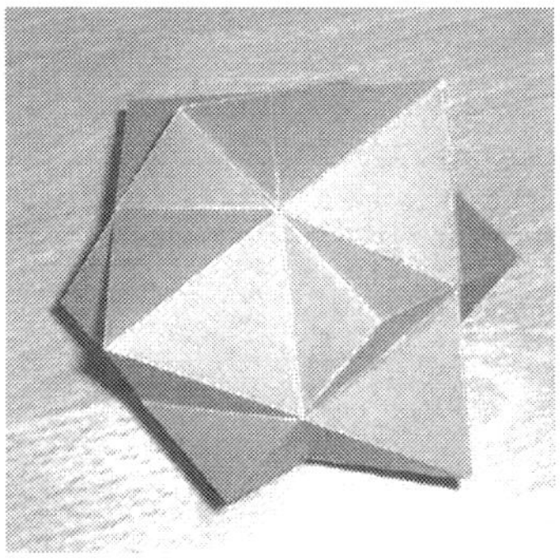

Cuboctahedron

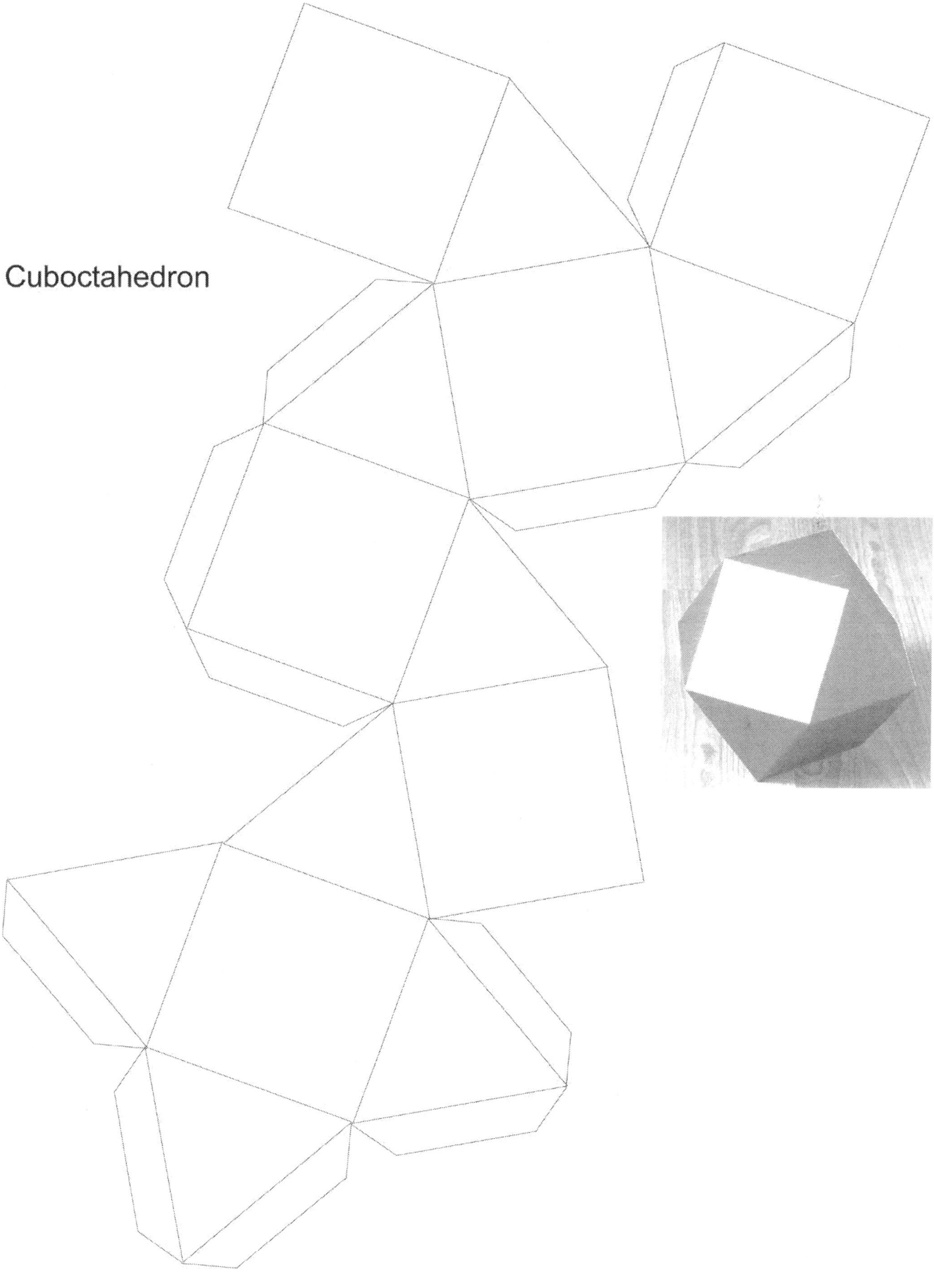

Dodecahedron

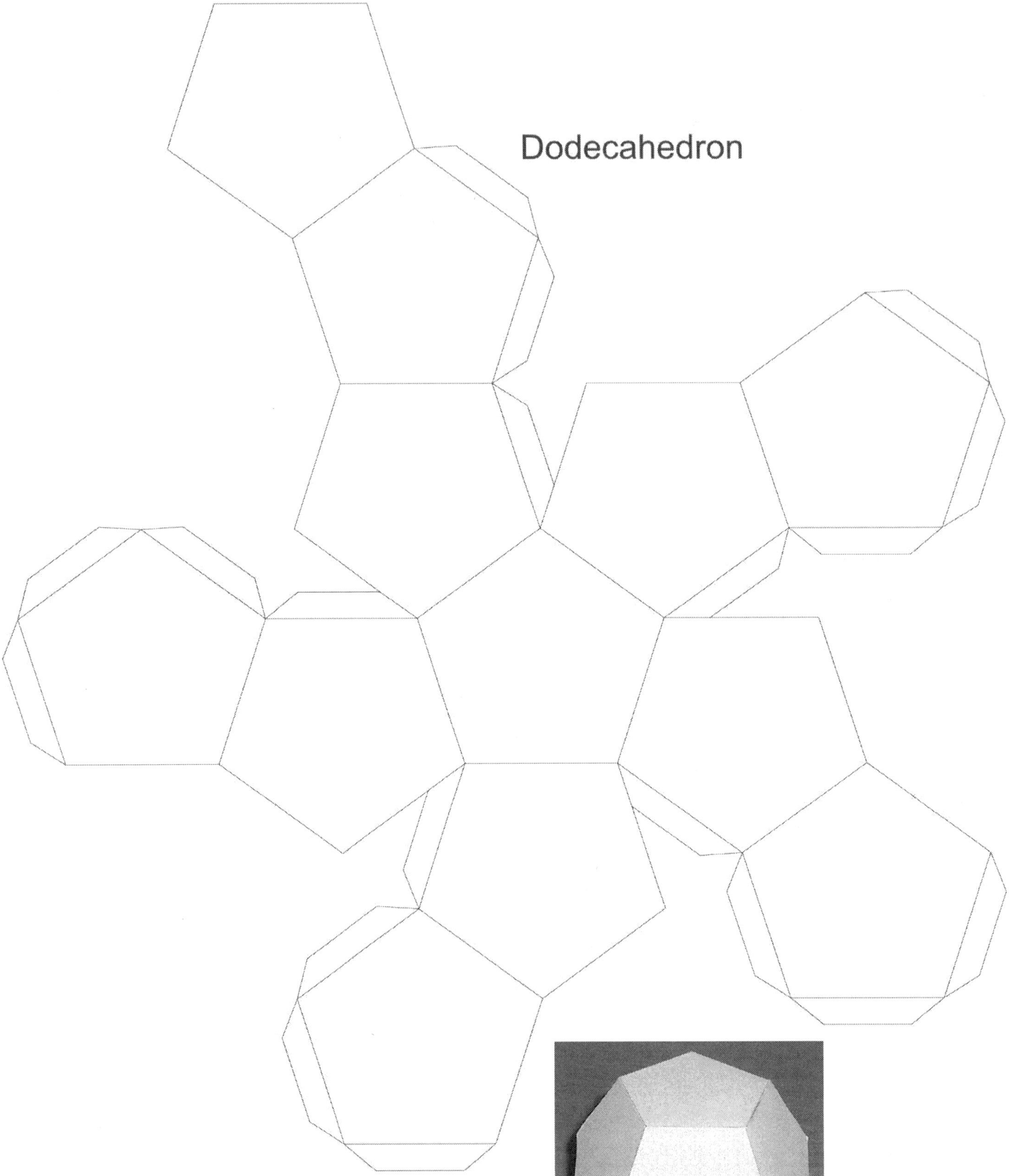

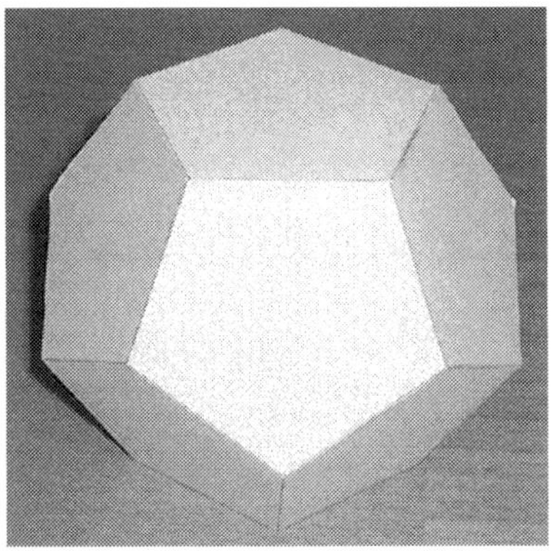

Great Stellated Dodecahedron
made out of one piece of paper.
Cut the lines between the long
and the short sides of the triangles.
Fold the long lines backwards and
fold the short lines forwards.

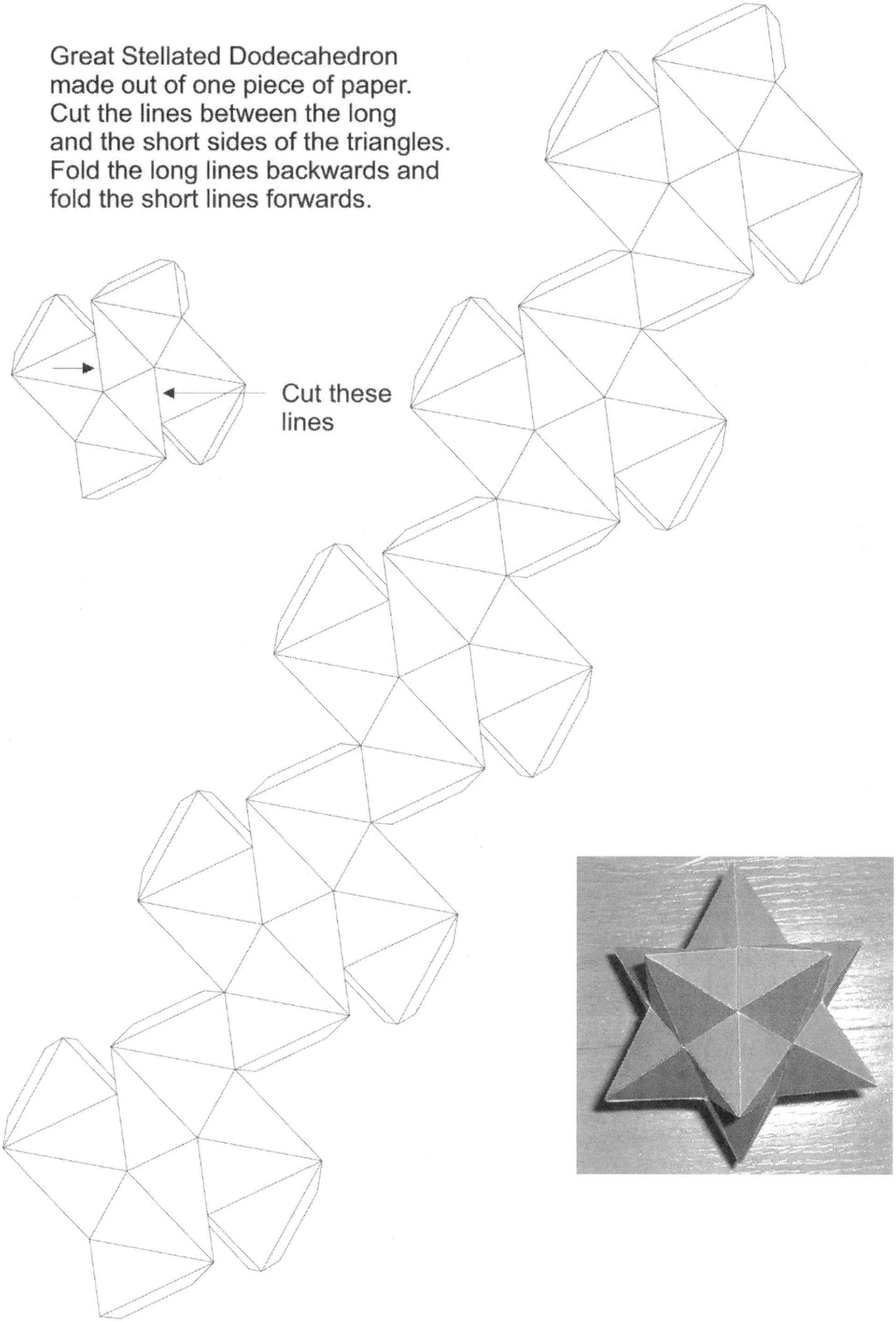

Cut these
lines

Hebdomicontadissaedron

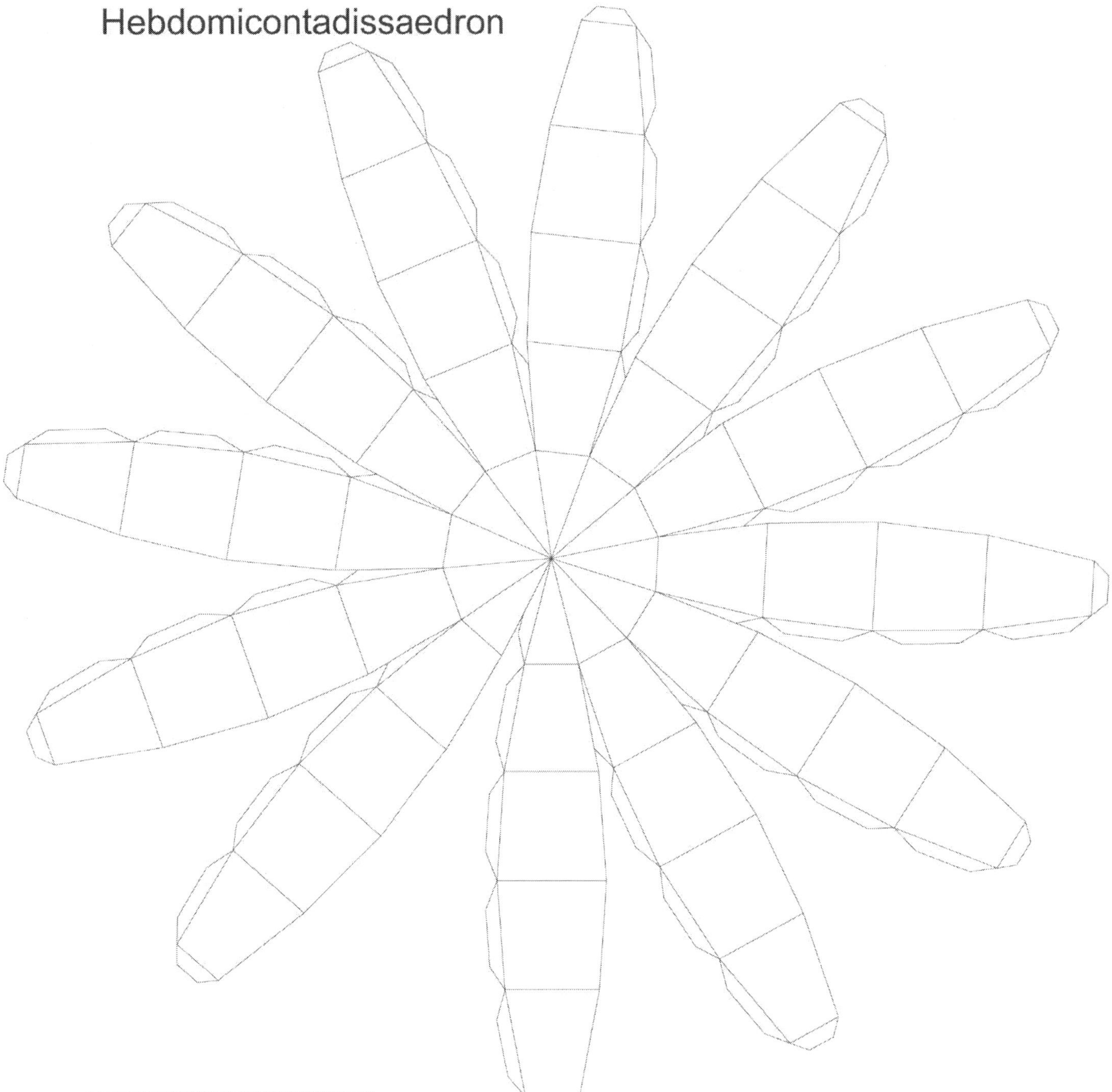

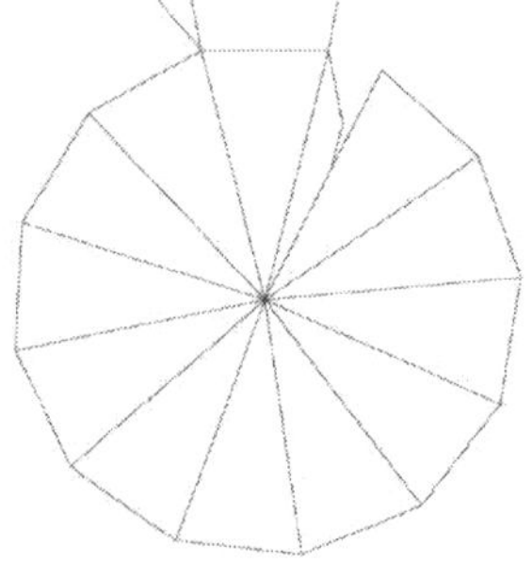

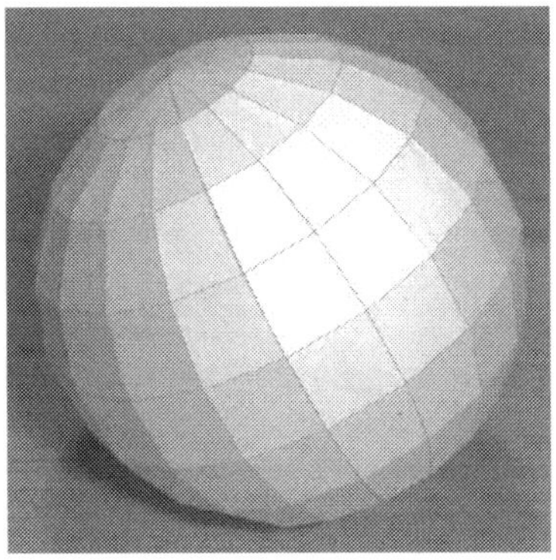

Truncated Hexagonal Prism

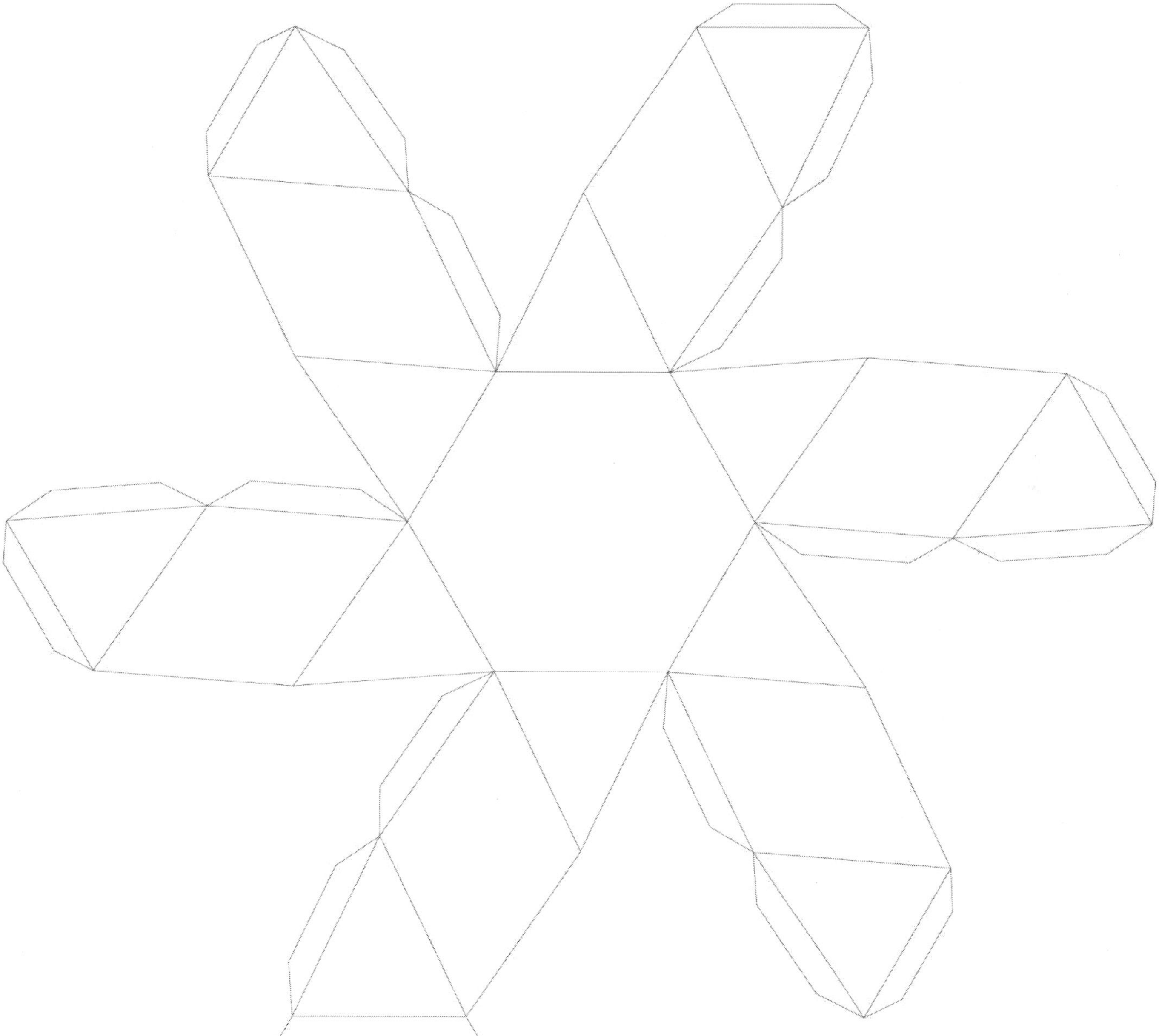

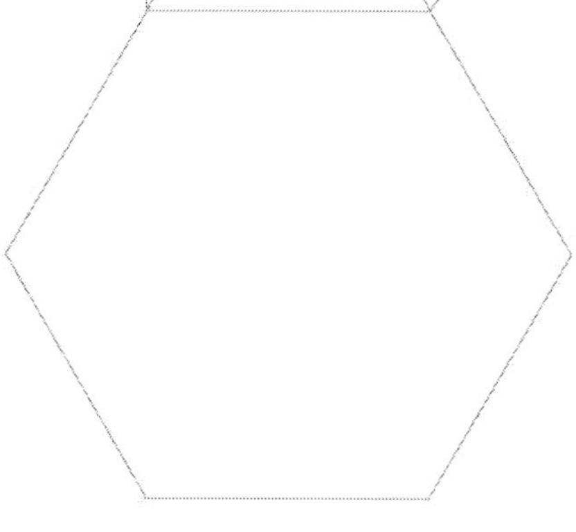

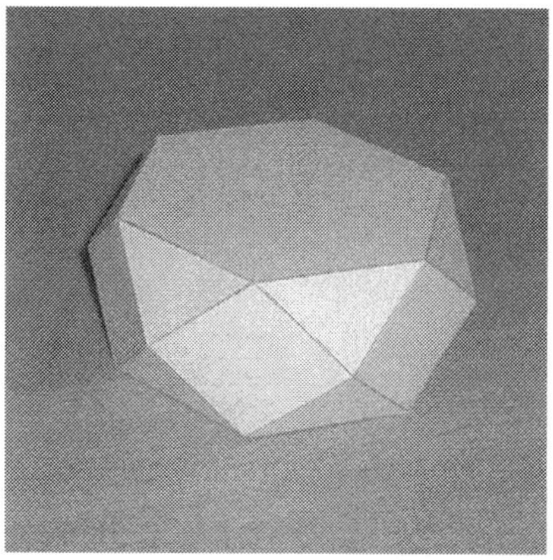

Hexagonal Antiprism

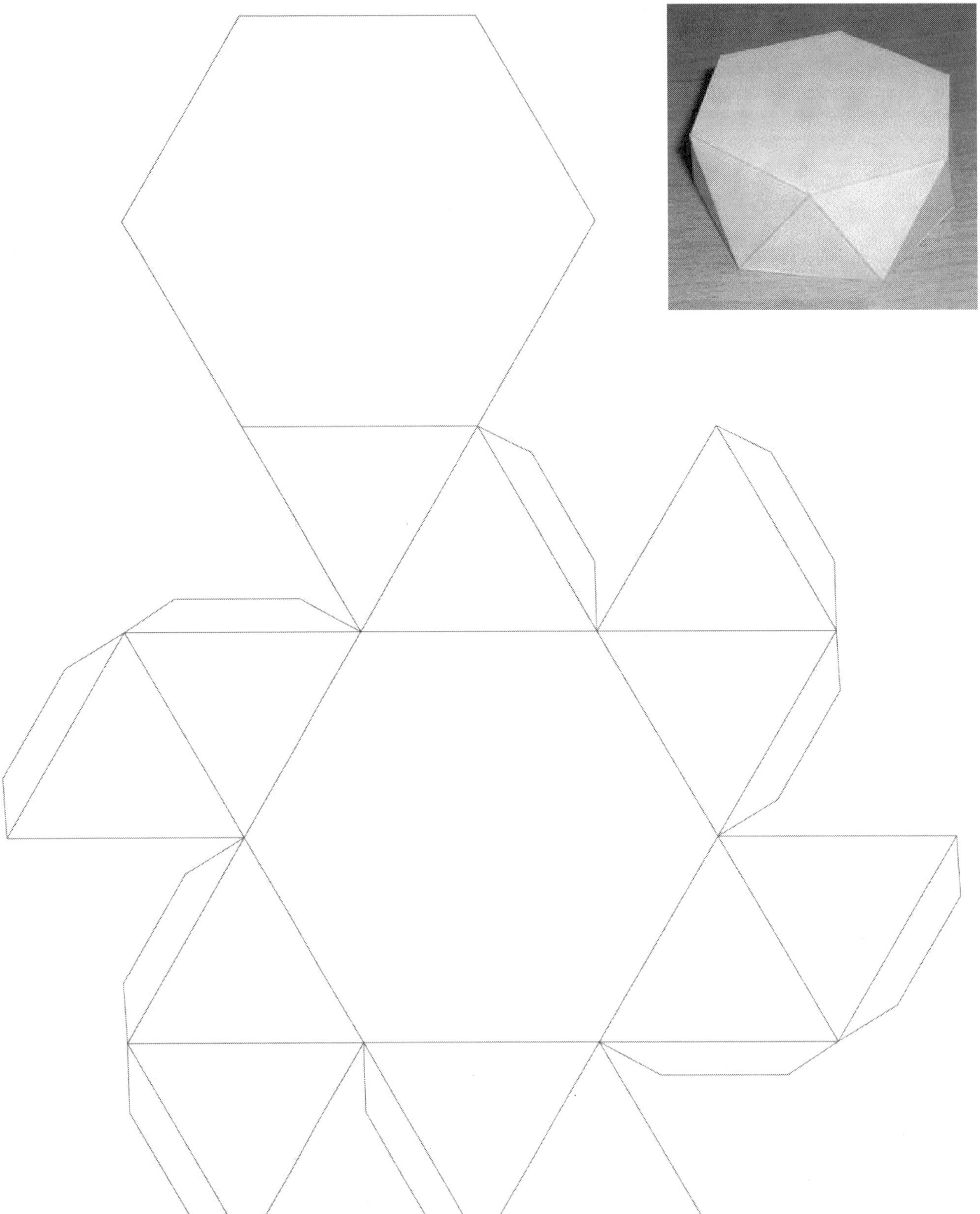

Hexagonal Prism

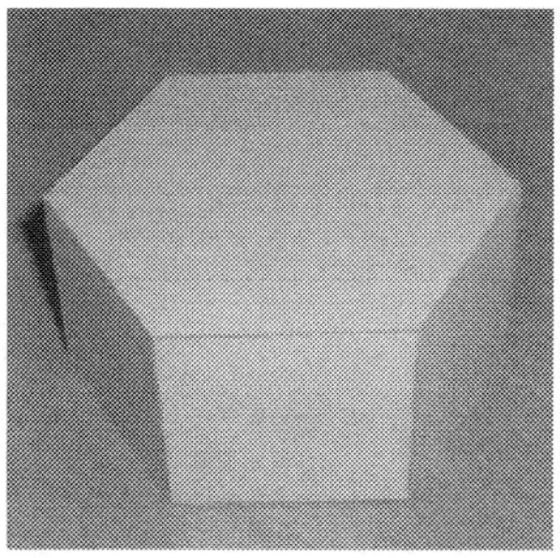

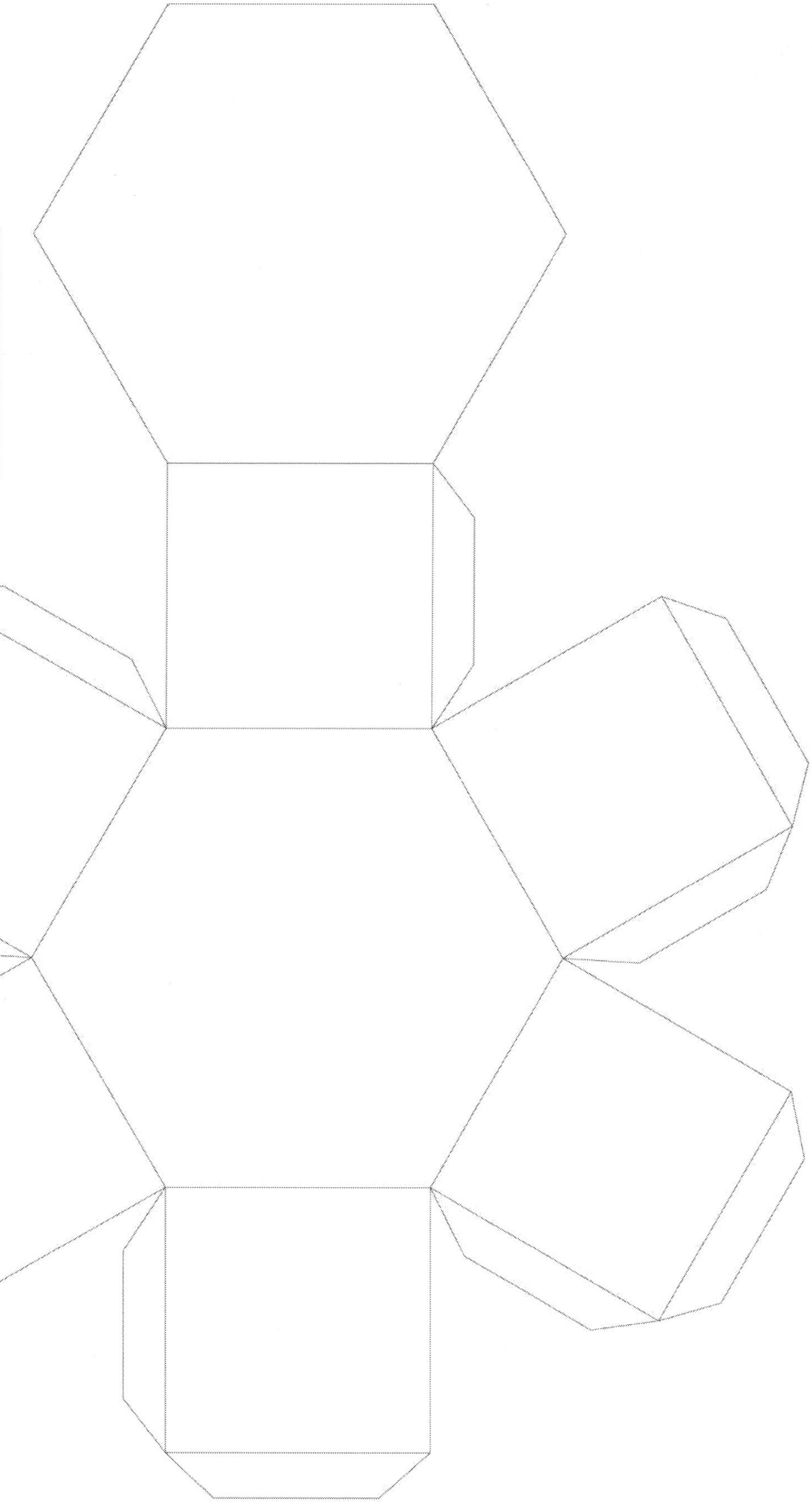

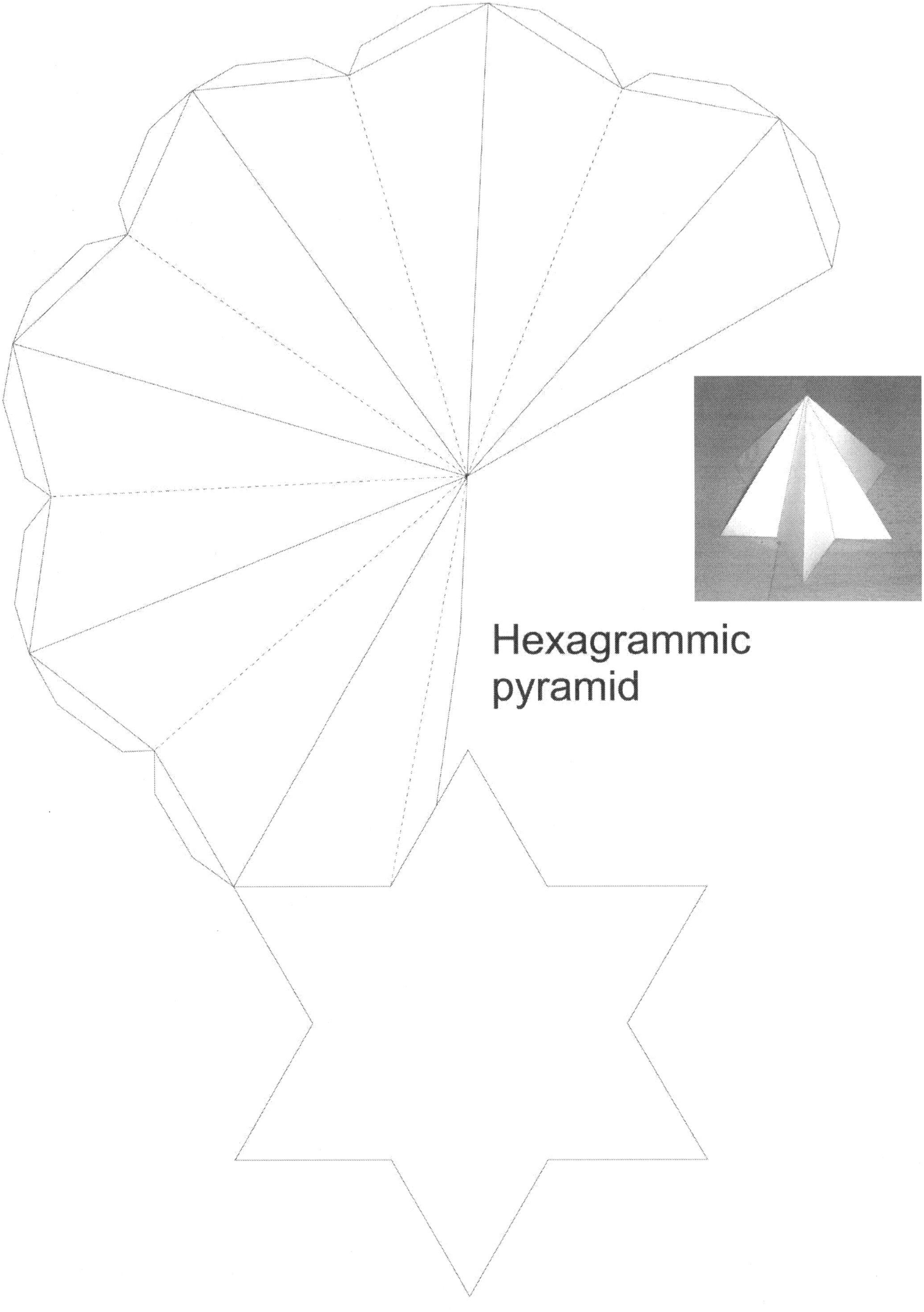

Hexagrammic pyramid

Hexagrammic Antiprism

Fold the dotted lines forwards
Fold the other lines backwards

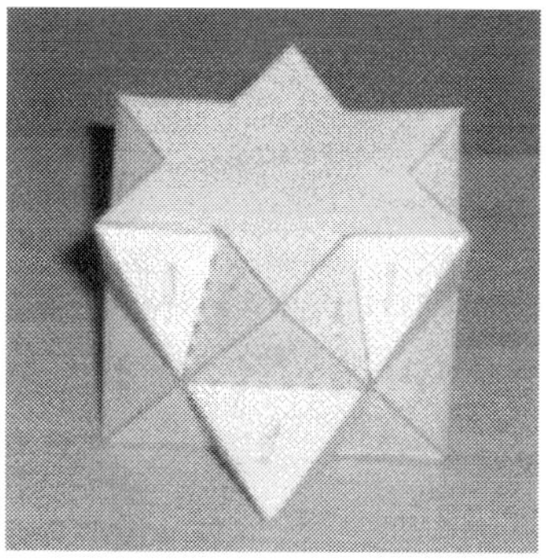

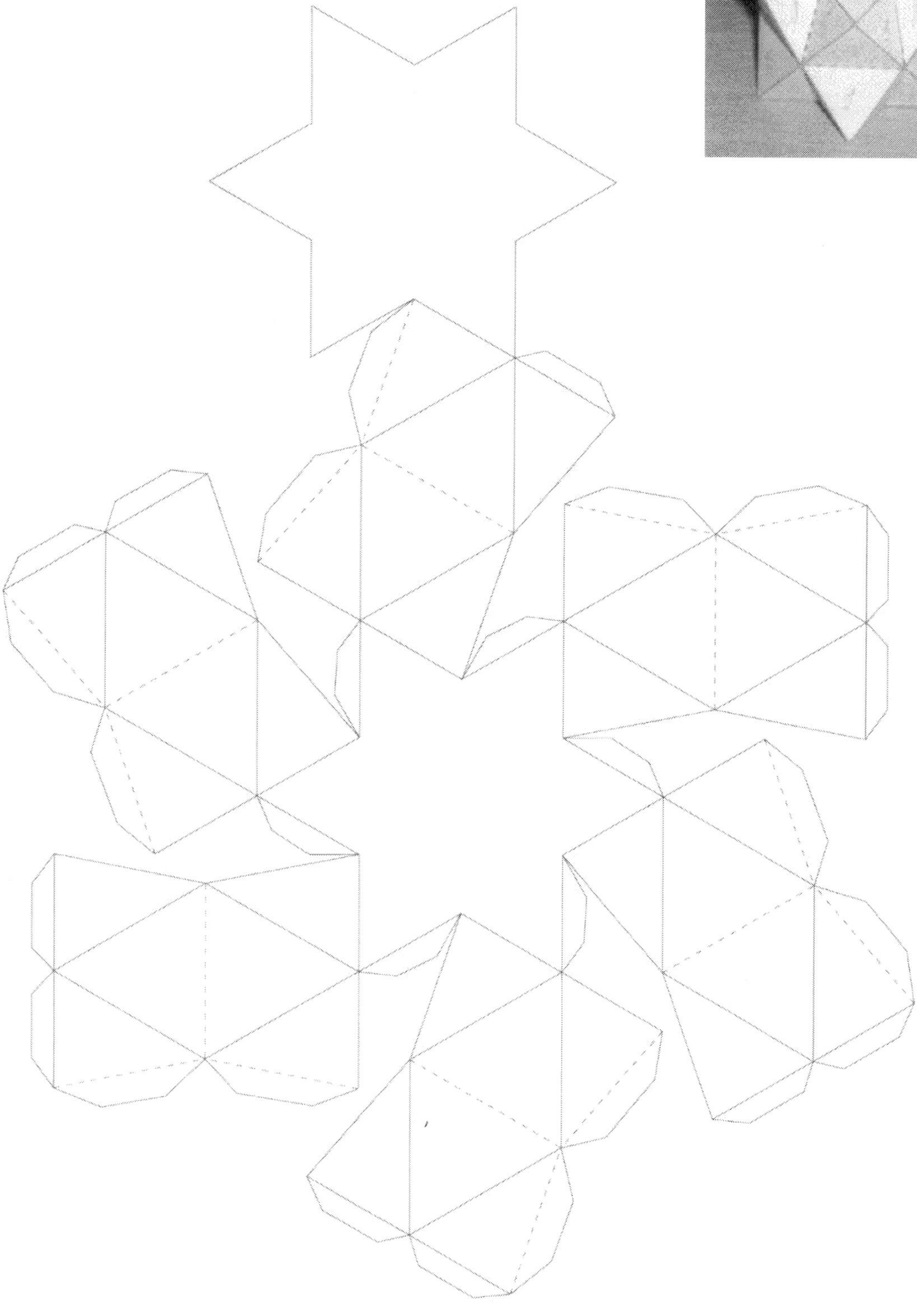

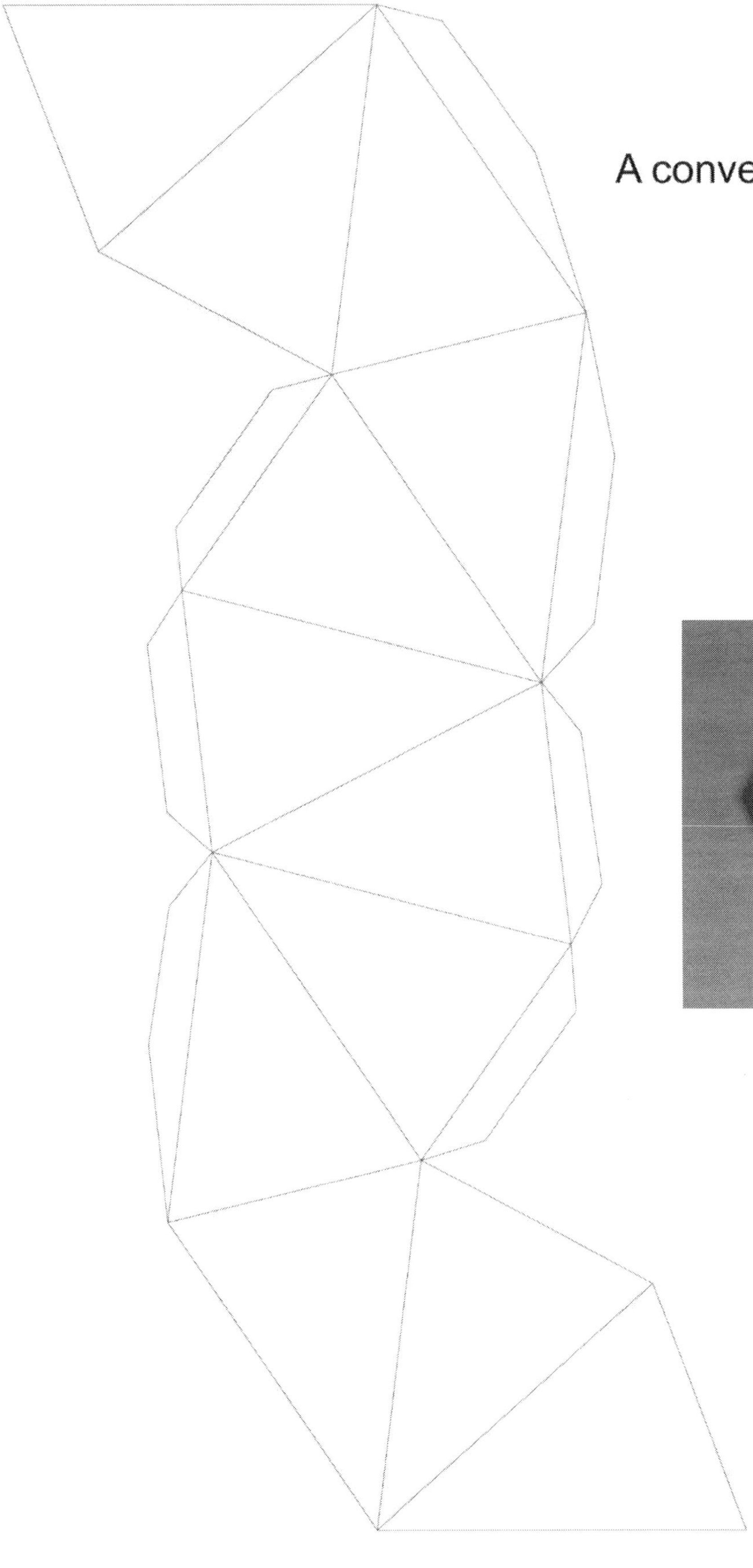

A convex dodecahedron

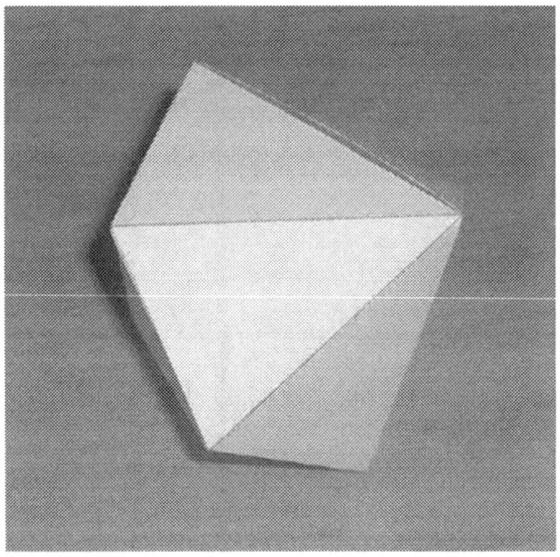

Octagonal Prism

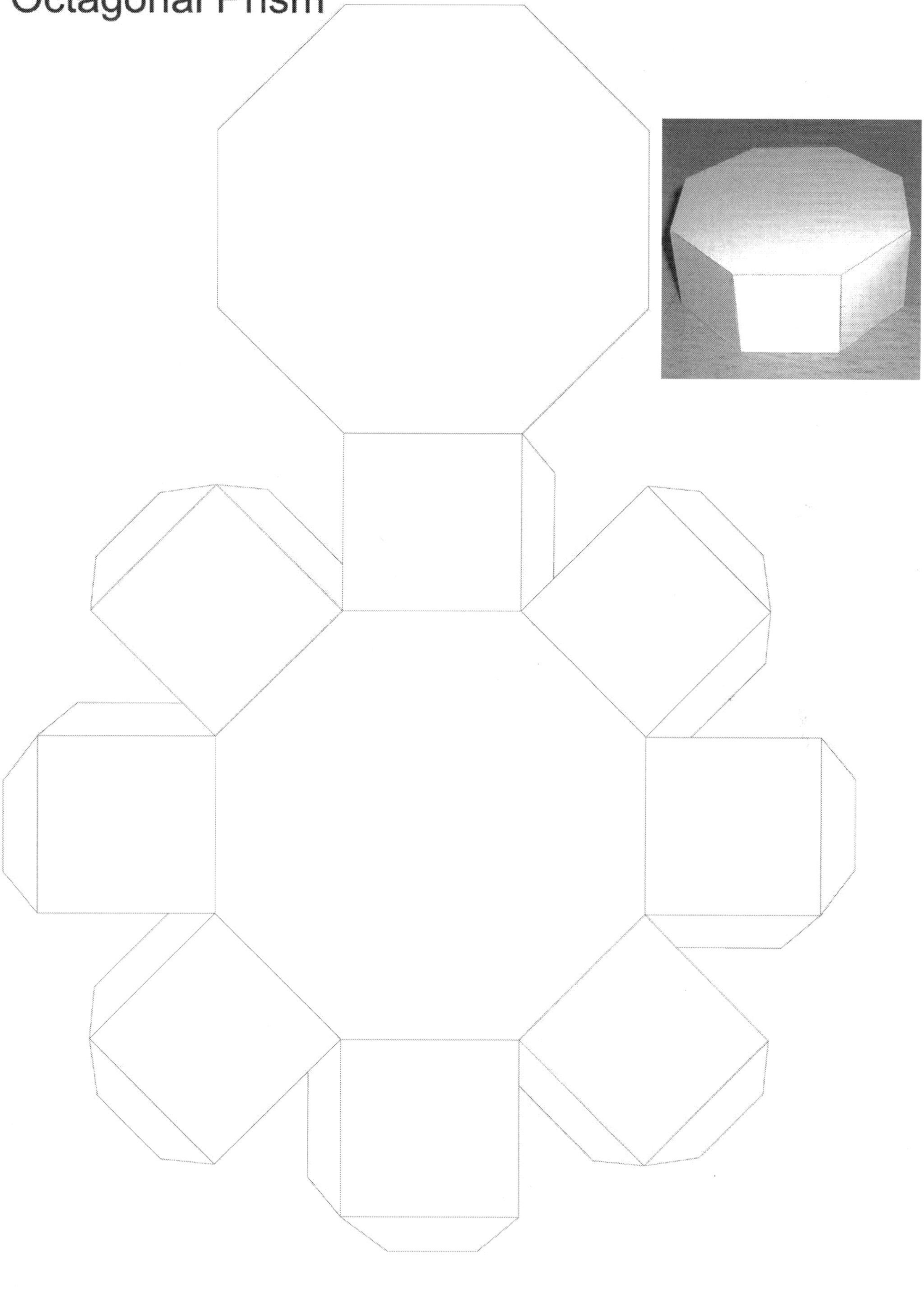

Pentagrammic Pyramid (
Fold the dotted lines forward
Fold the other lines backwards

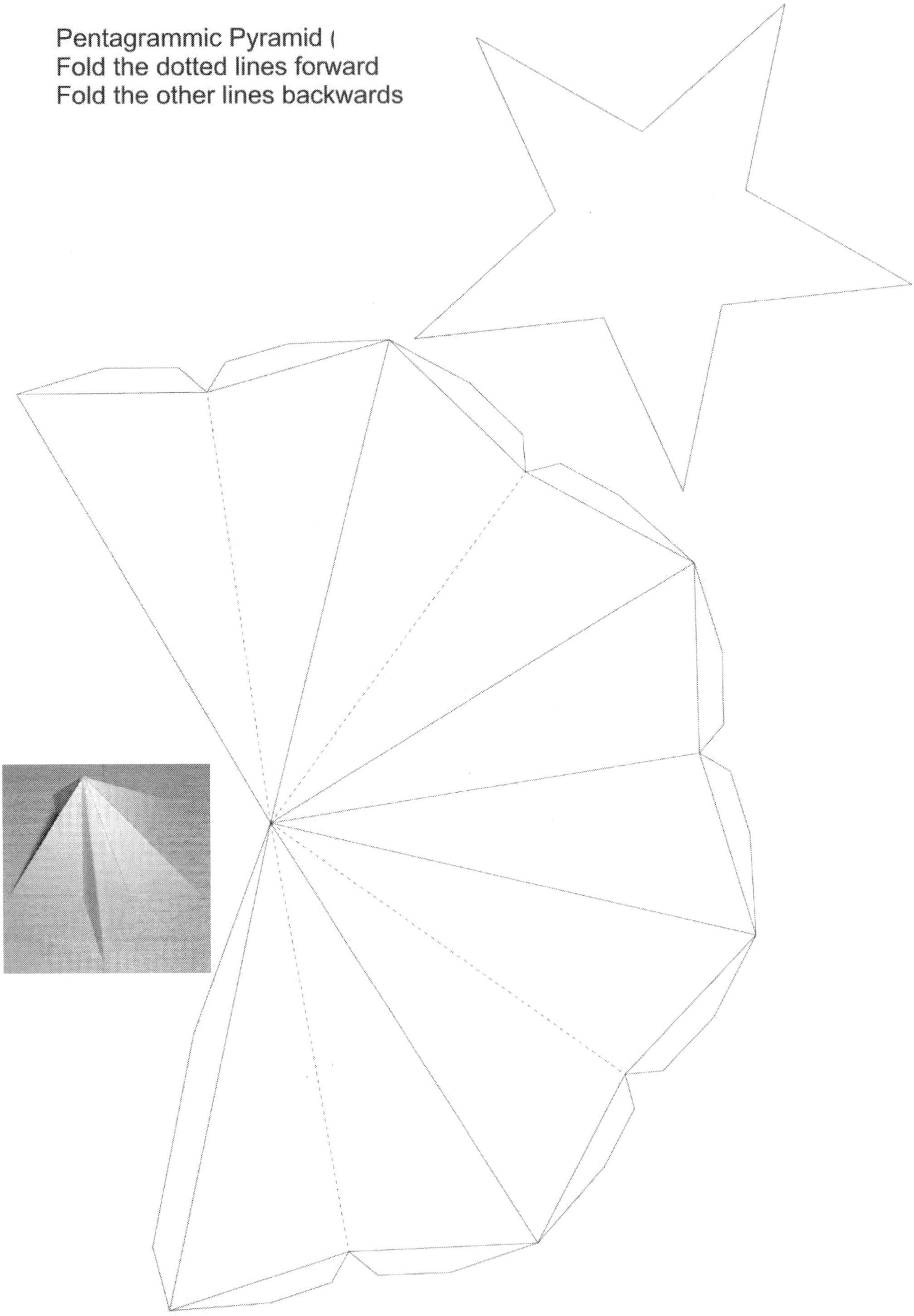

Pentagrammic Antiprism

Fold the dotted lines forwards
Fold the other lines backwards

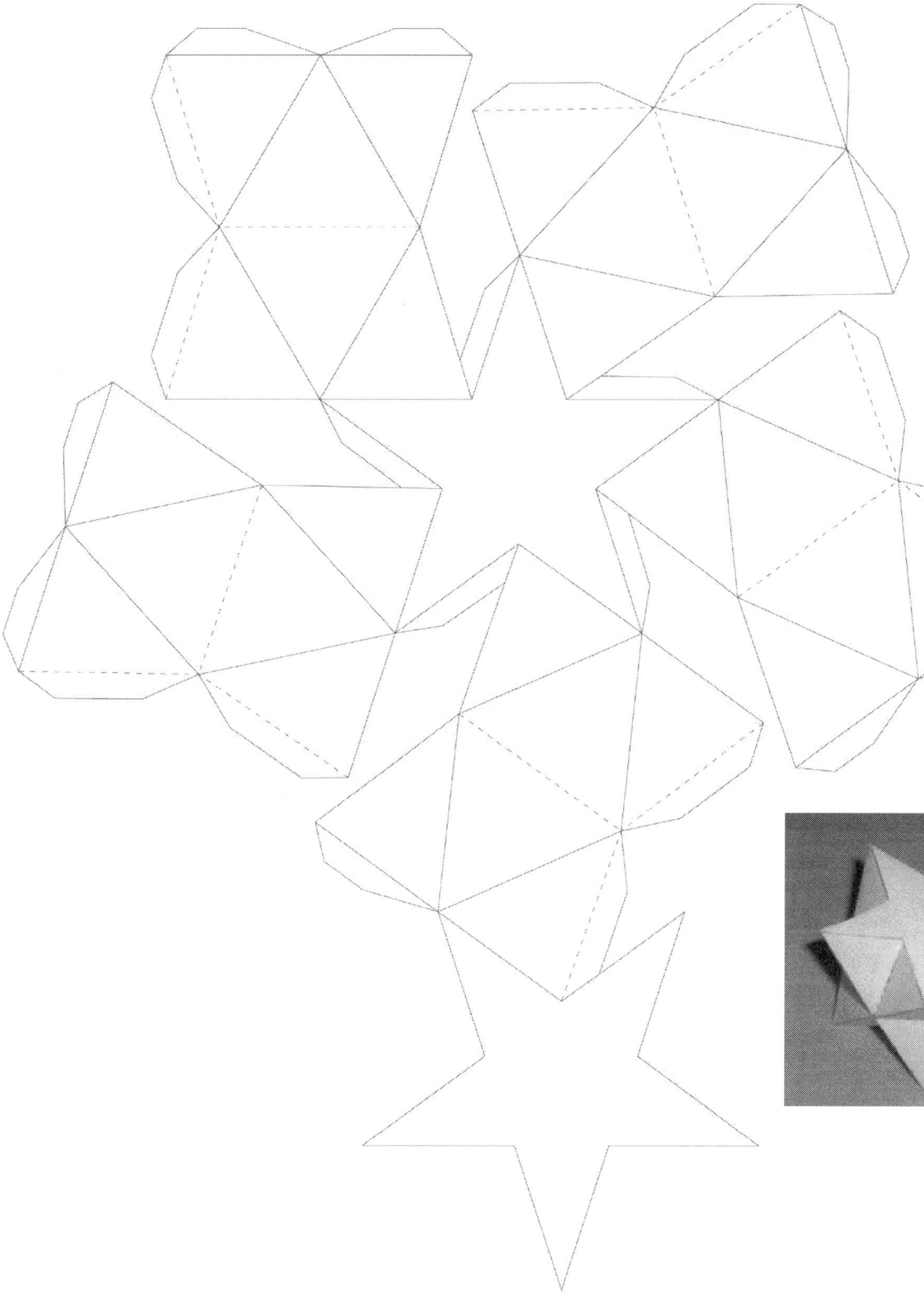

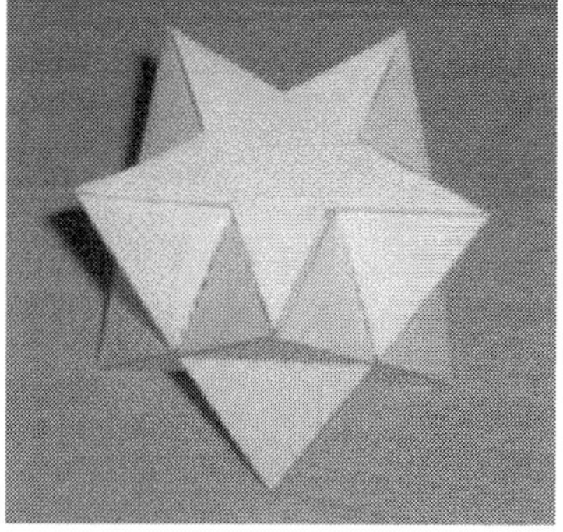

Pentagrammic Prism

Fold the dotted lines forwards
Fold the other lines backwards

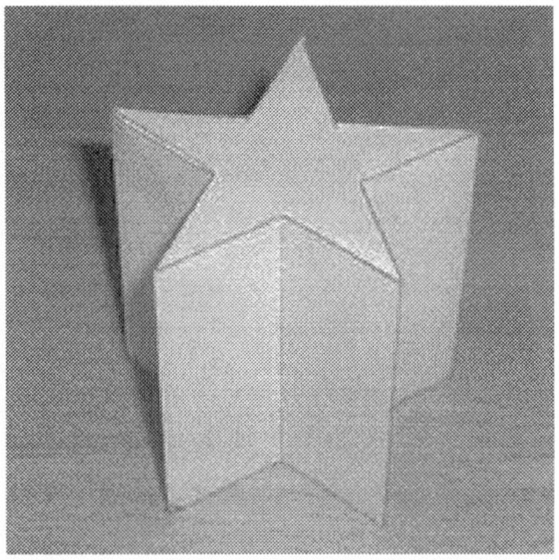

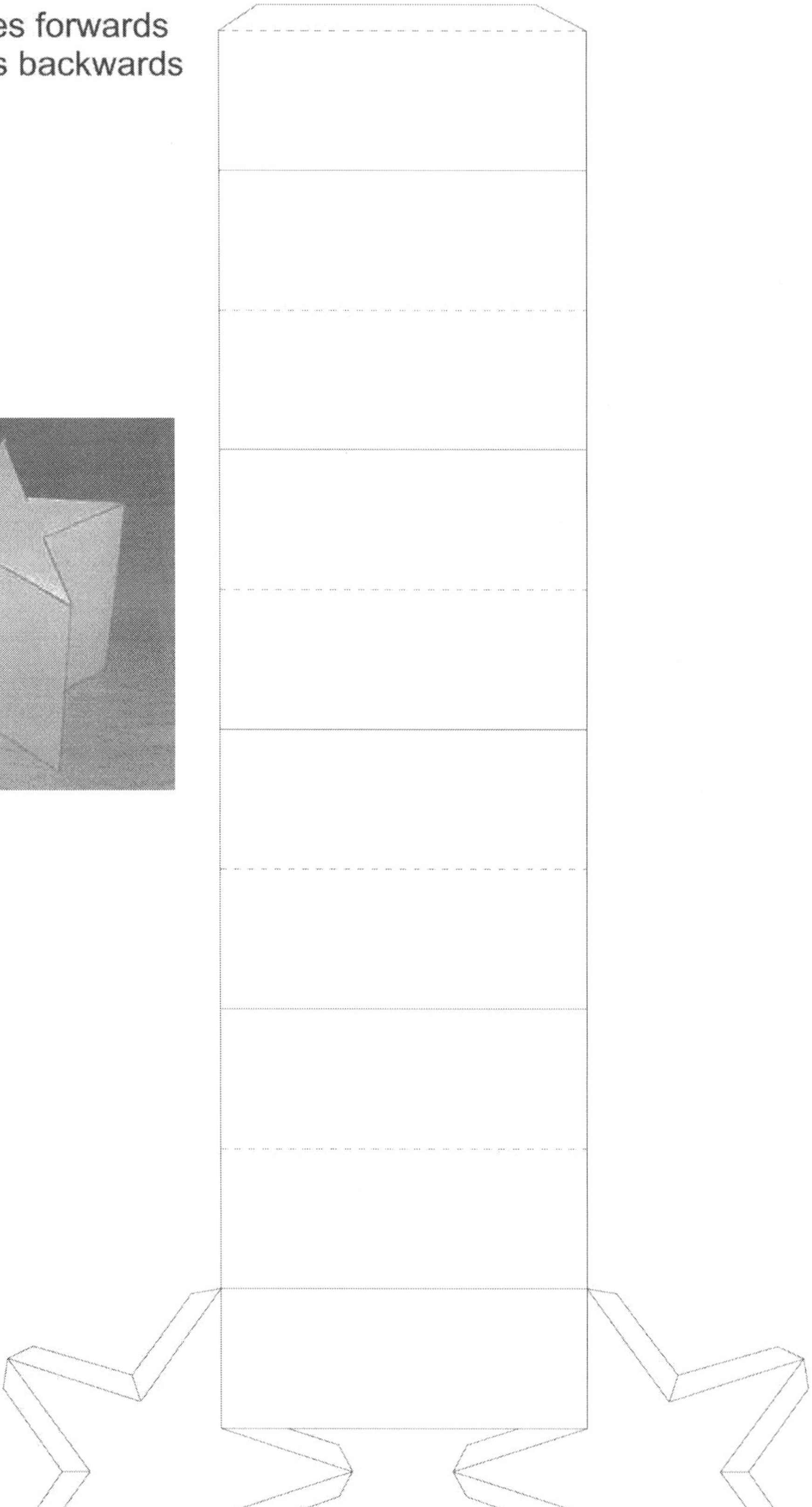

Rhombic Dodecahedron

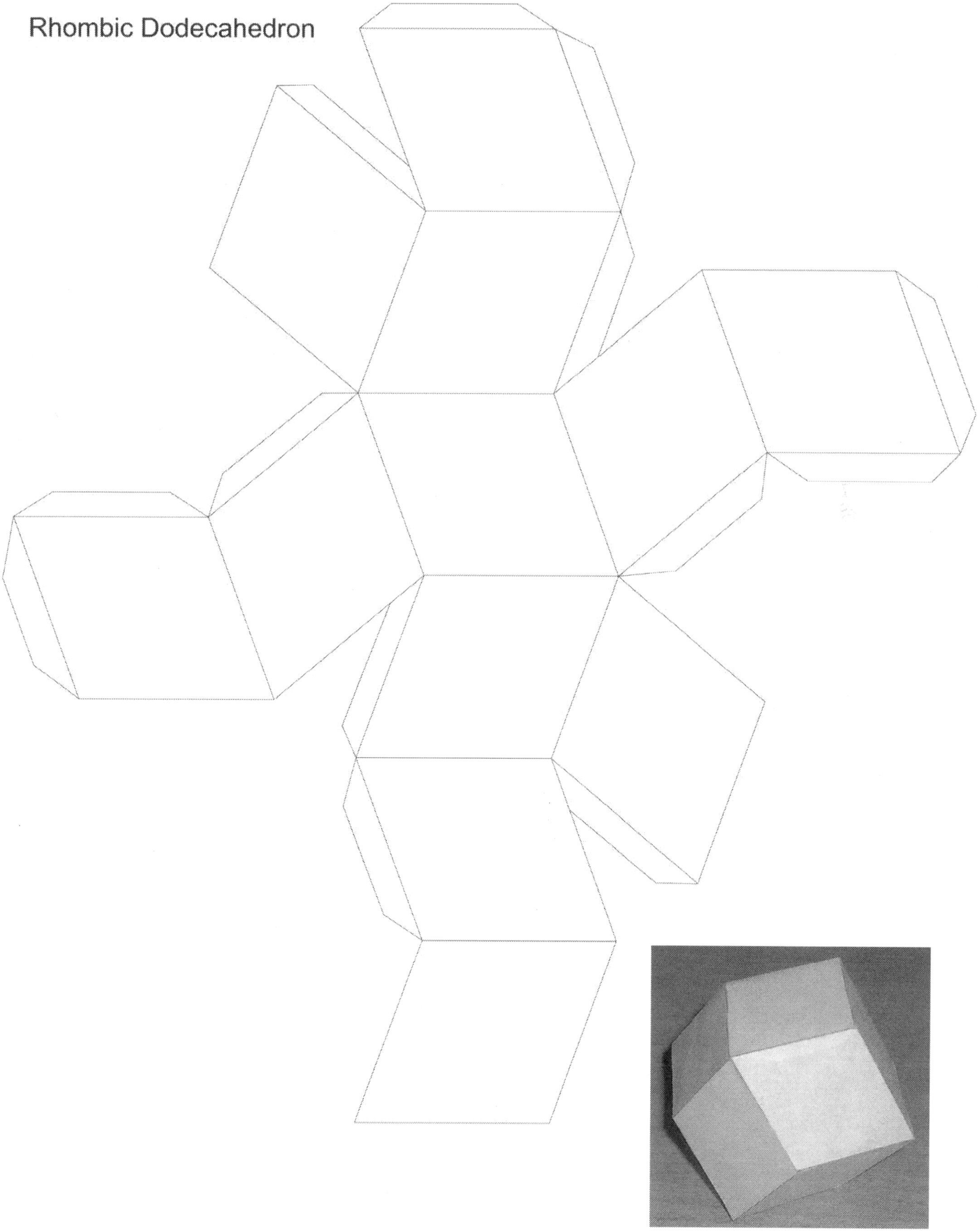

Rhombicosidodecahedron

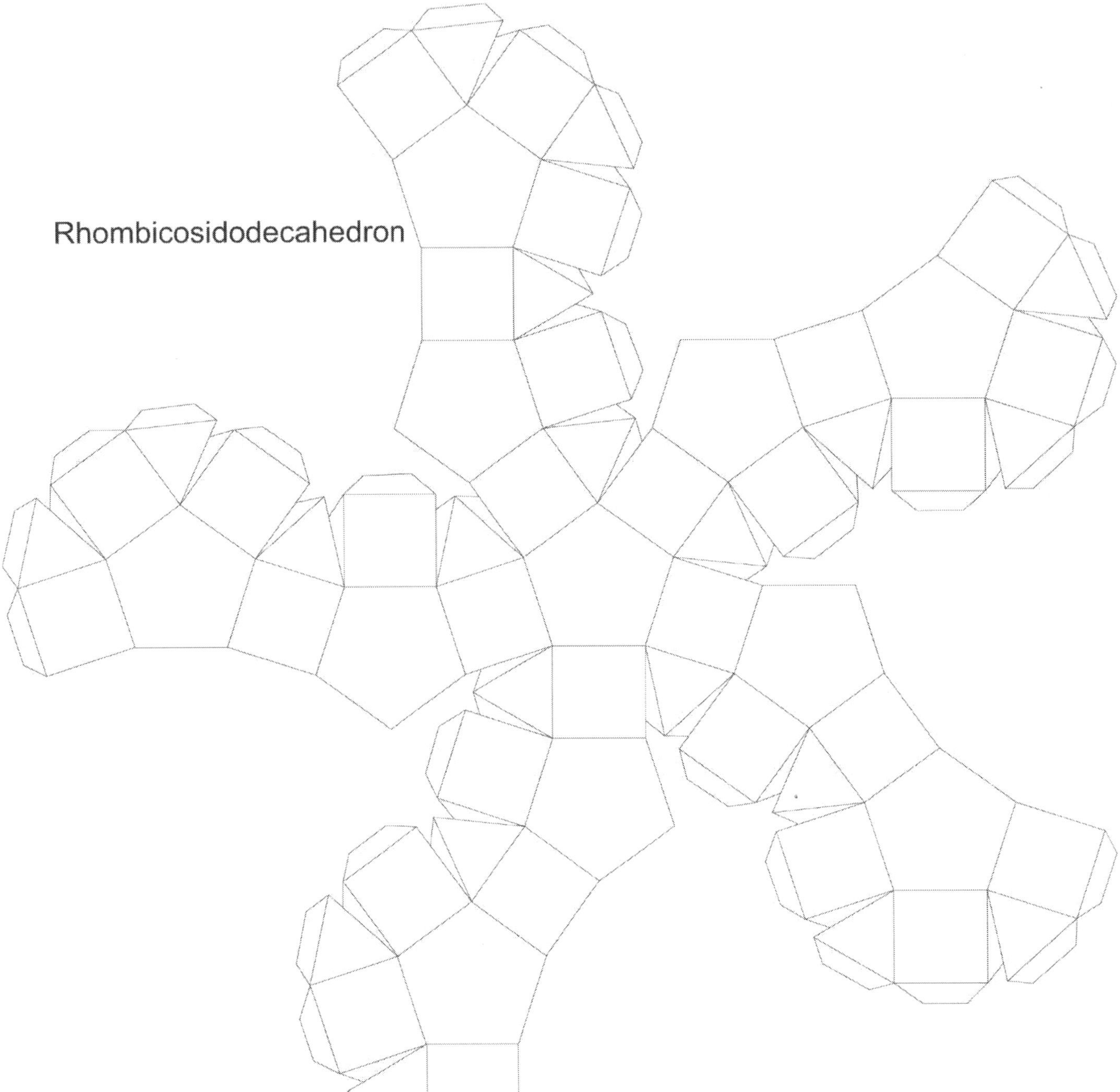

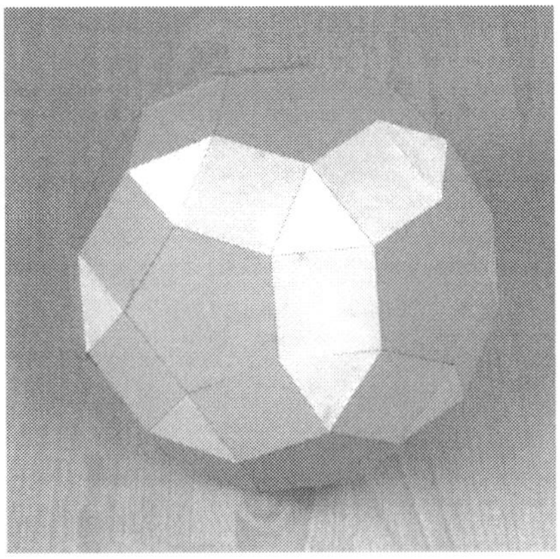

Stellated Dodecahedron

Fold the long lines backwards
Fold the short lines forwards

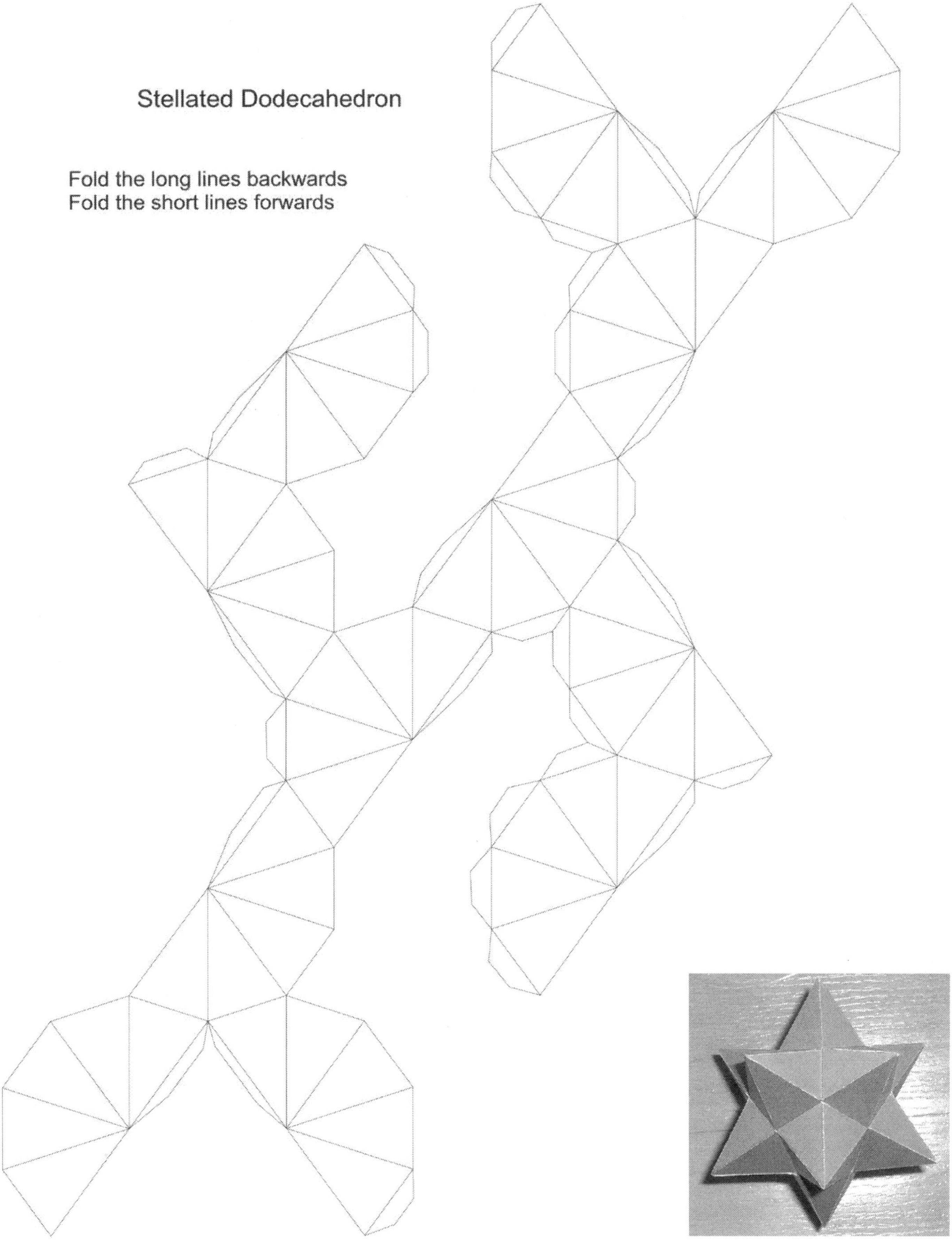

Triambic Icosahedron

Fold the dotted lines forwards
Fold the other lines backwards

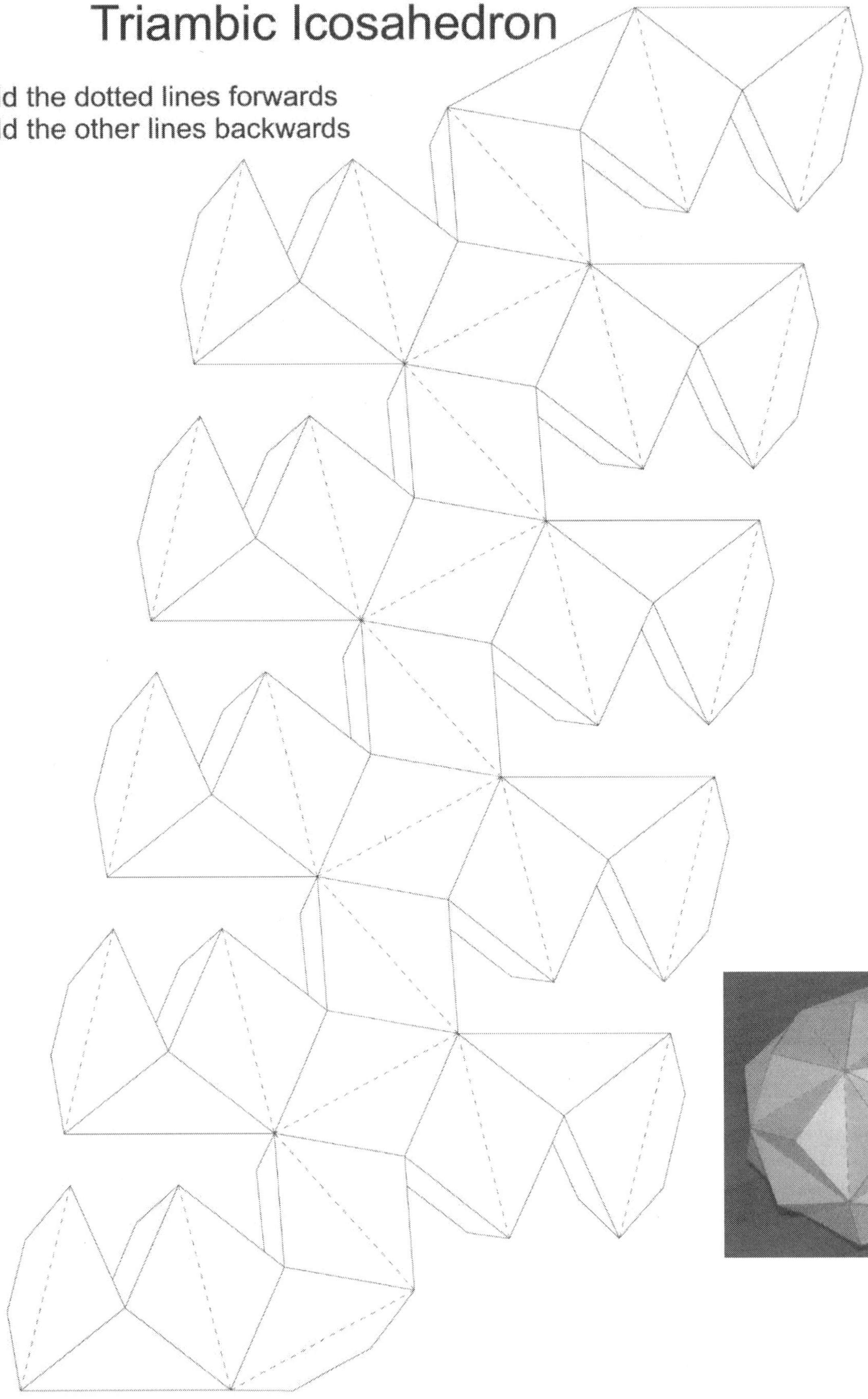

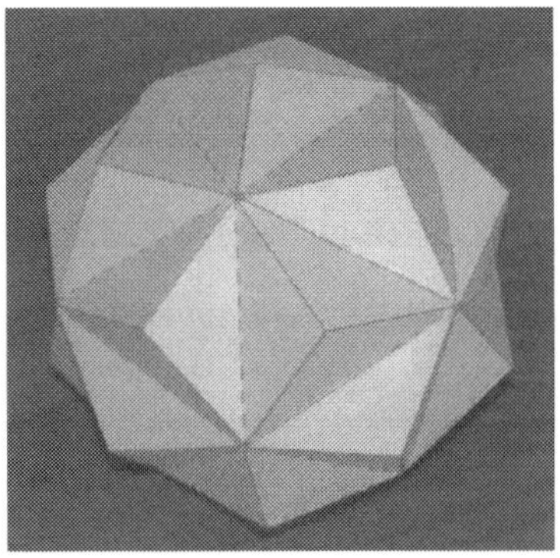

Snub cube

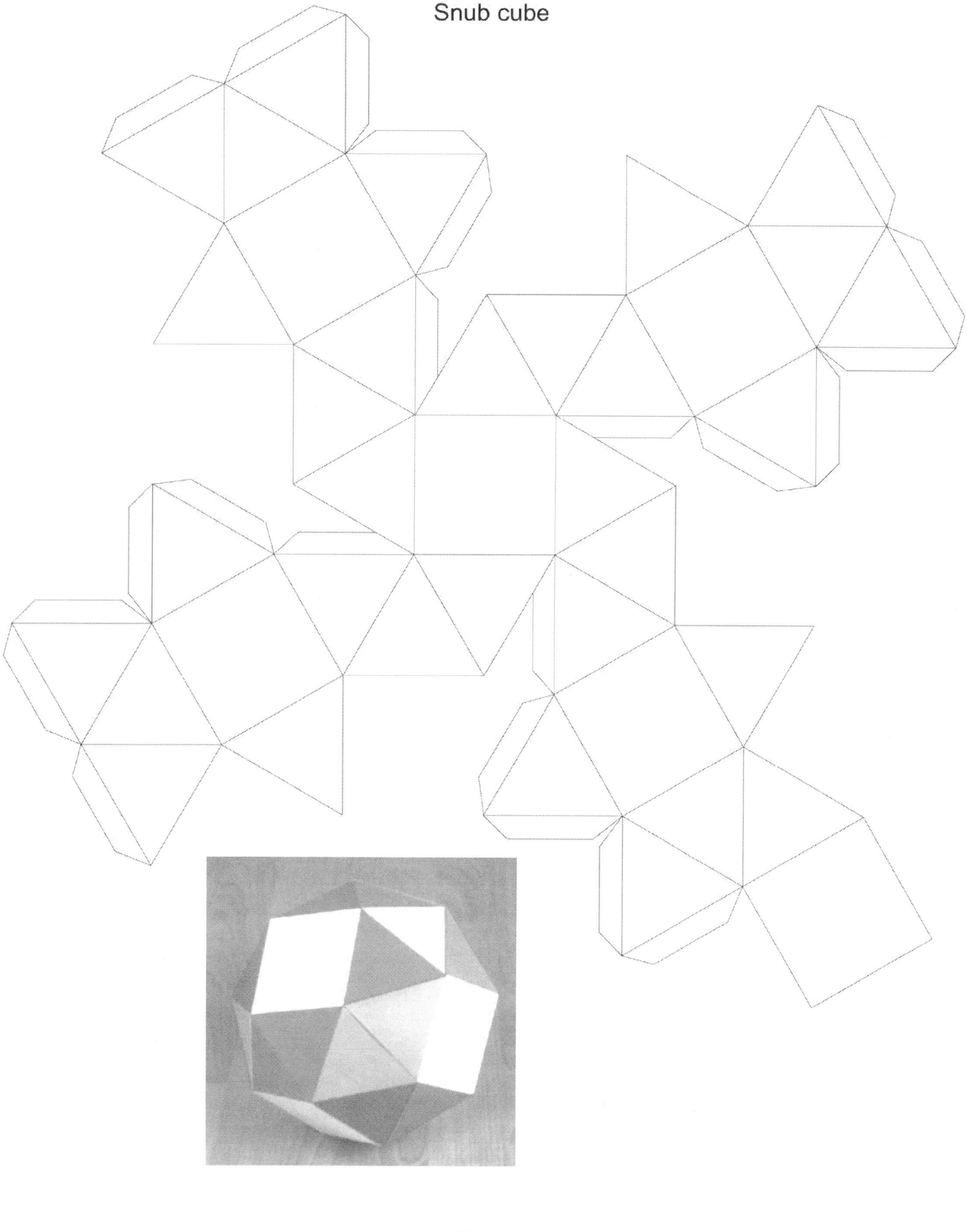

Truncated cuboctahedron

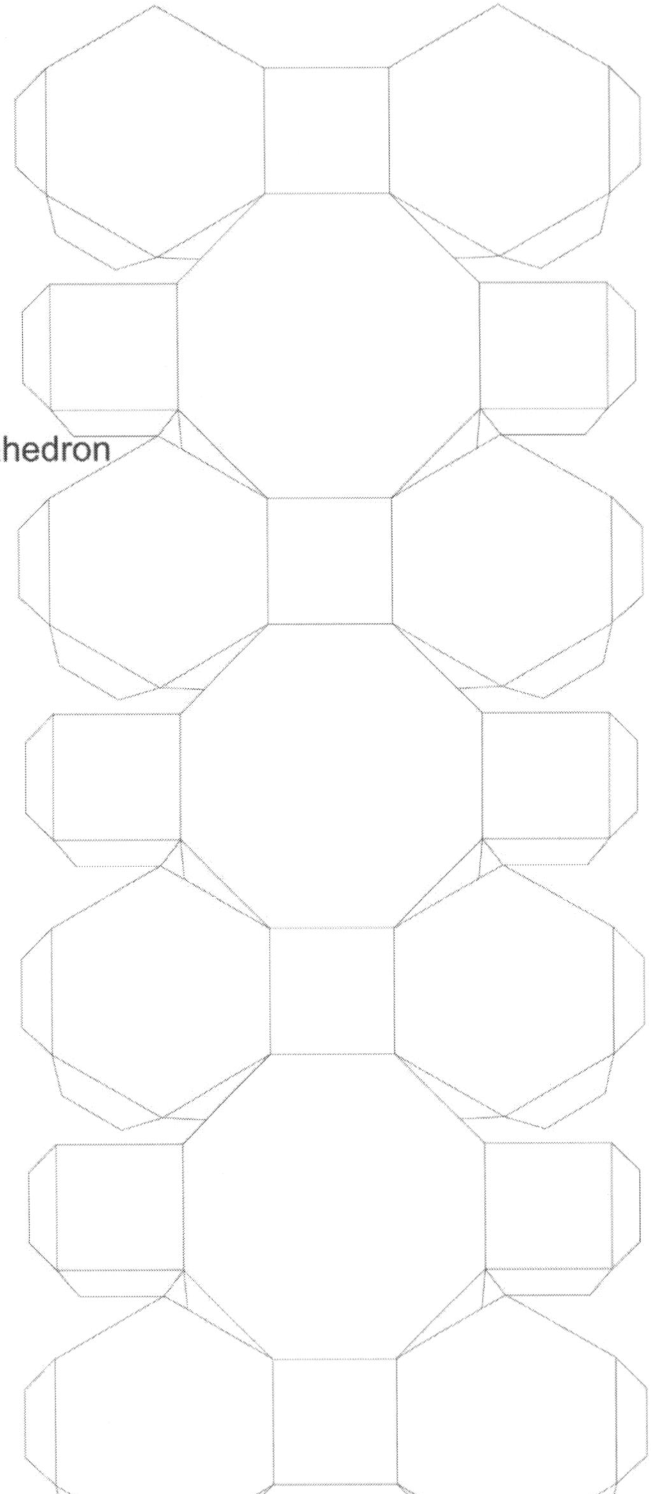

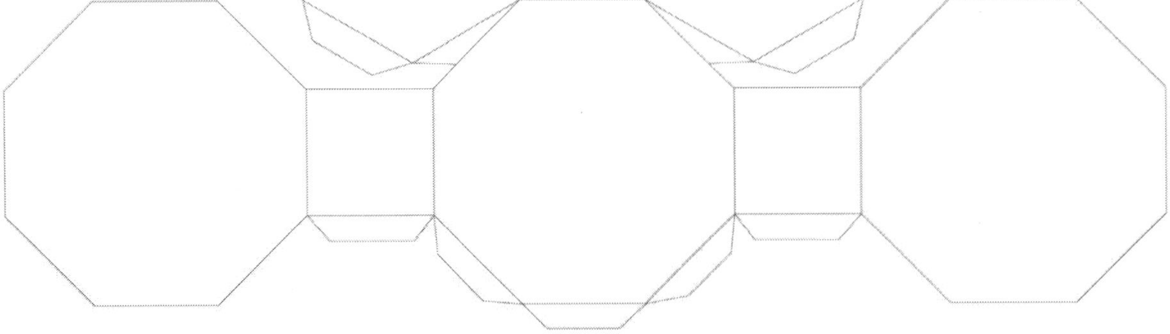

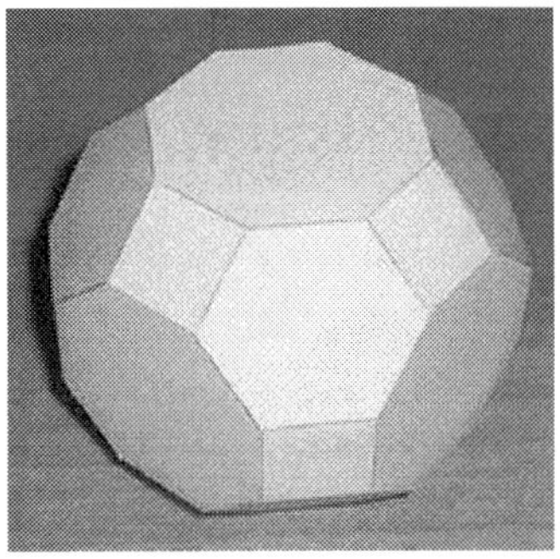

Truncated Dodecahedron

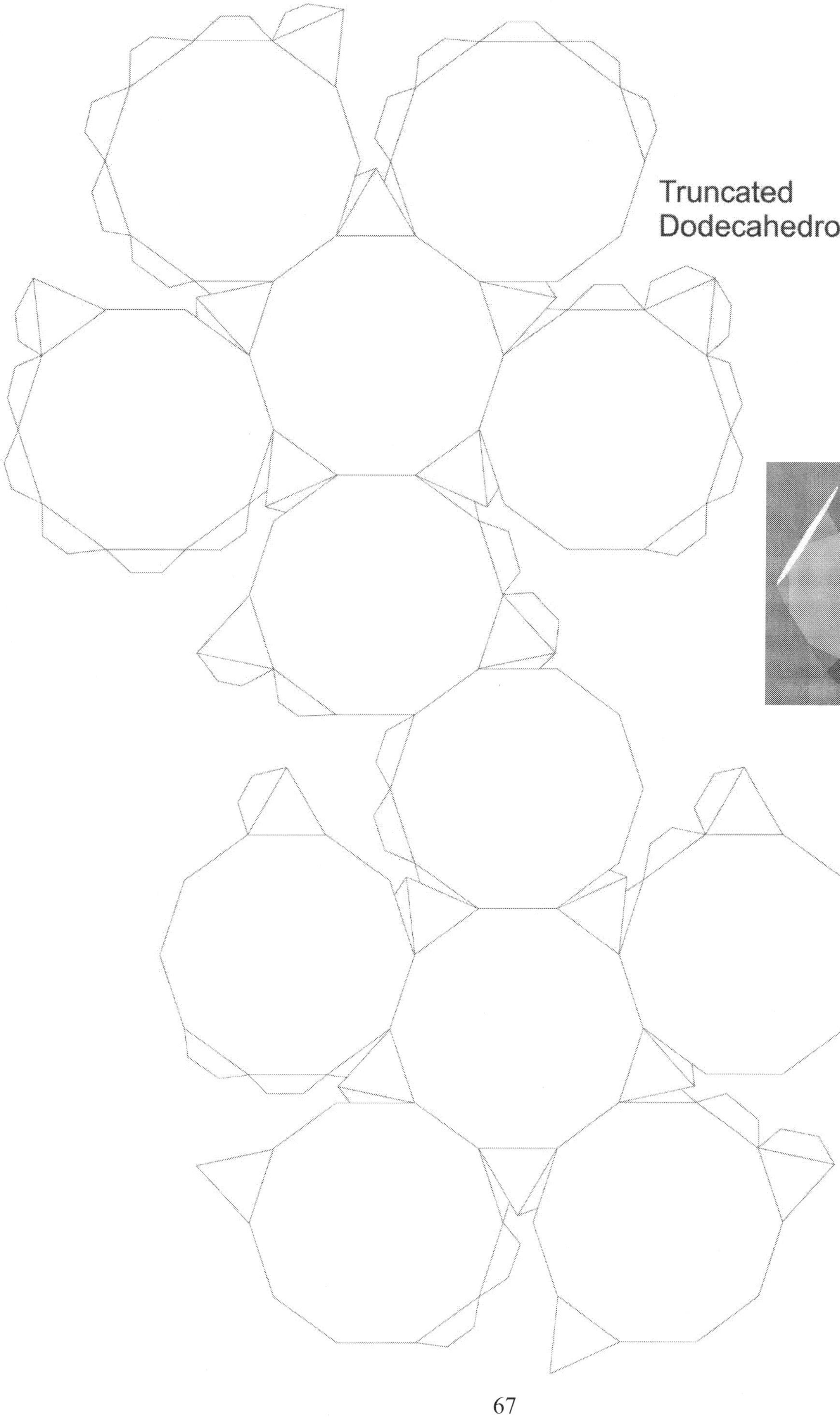

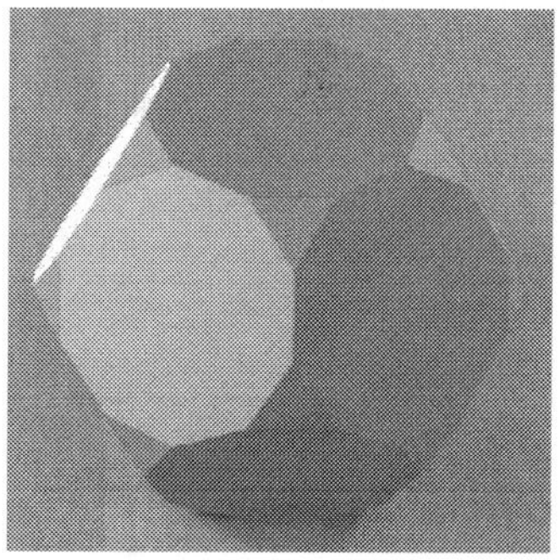

Truncated Icosahedron

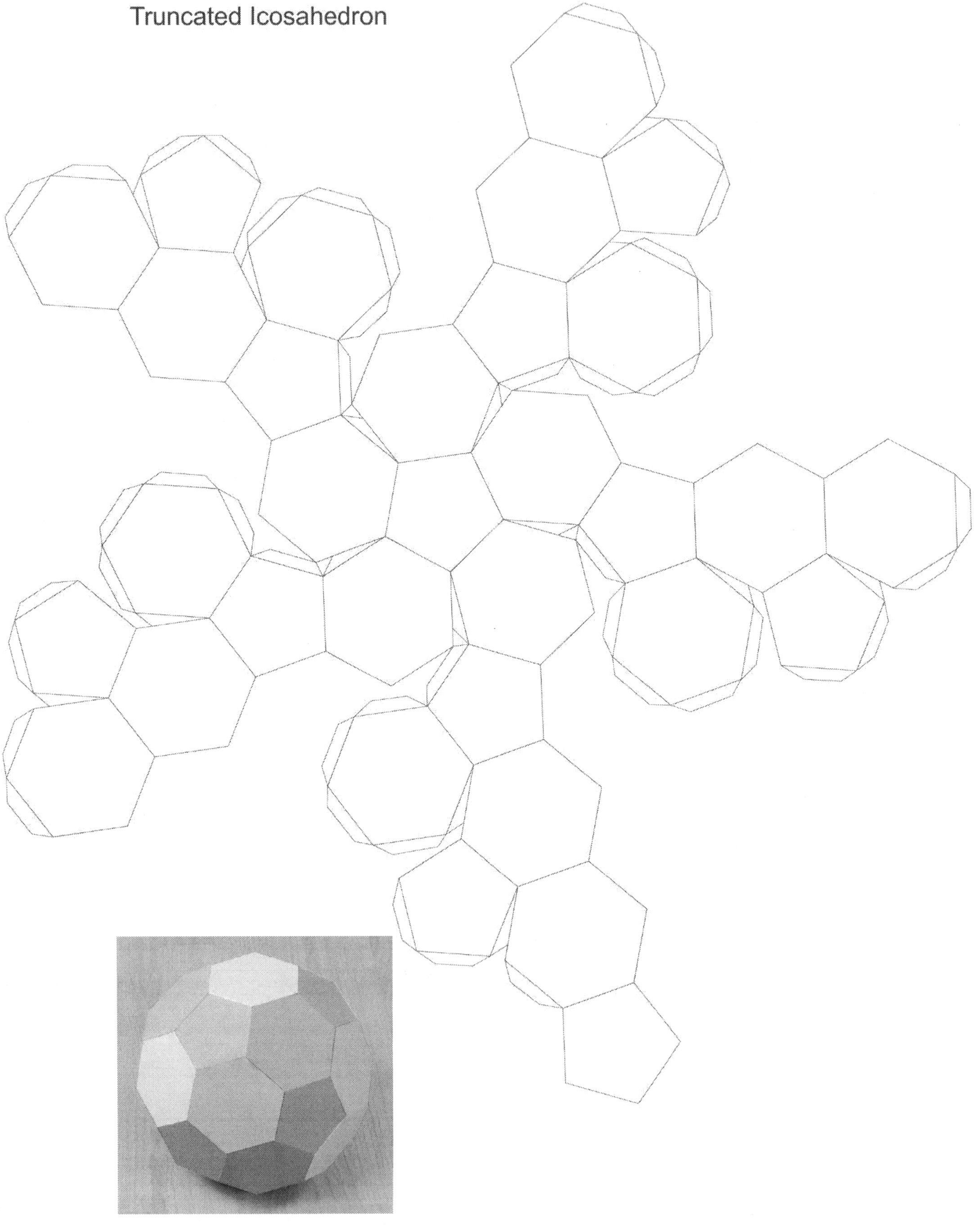

Truncated Octahedron

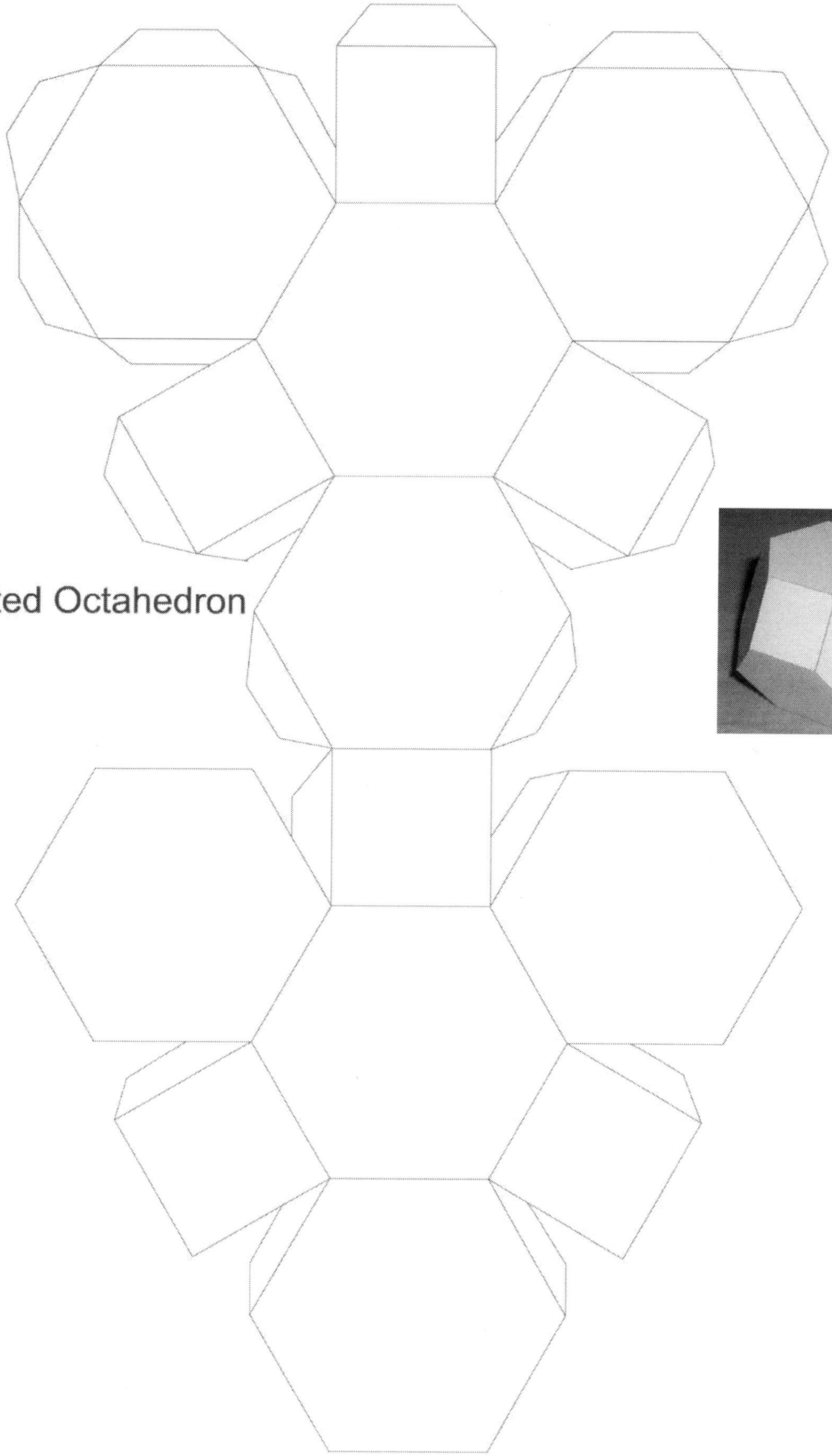

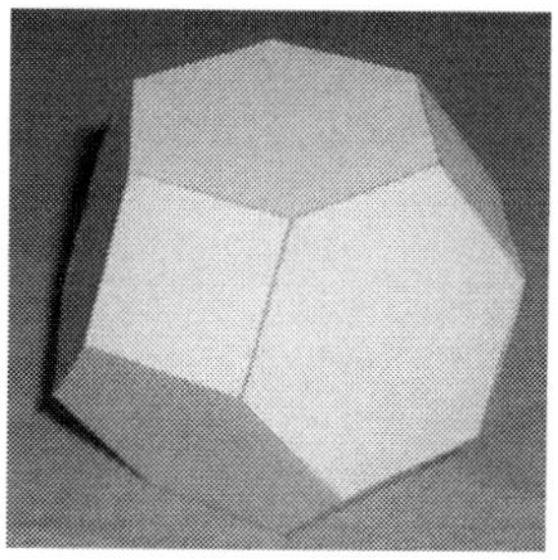